SCIENTIFIC PAPERS AND
ADDRESSES OF
THE HON. SIR CHARLES A. PARSONS

PLATE I

THE HON. SIR CHARLES ALGERNON PARSONS, O.M., K.C.B., F.R.S.

SCIENTIFIC PAPERS AND ADDRESSES
OF
THE HON.
SIR CHARLES A. PARSONS
O.M., K.C.B., F.R.S.

With a Memoir by
LORD RAYLEIGH
and Appendices

EDITED BY
THE HON. G. L. PARSONS

CAMBRIDGE
AT THE UNIVERSITY PRESS
1934

CAMBRIDGE
UNIVERSITY PRESS

University Printing House, Cambridge CB2 8BS, United Kingdom

Cambridge University Press is part of the University of Cambridge.

It furthers the University's mission by disseminating knowledge in the pursuit of
education, learning and research at the highest international levels of excellence.

www.cambridge.org
Information on this title: www.cambridge.org/9781107502024

© Cambridge University Press 1934

First published 1934
First paperback edition 2015

A catalogue record for this publication is available from the British Library

ISBN 978-1-107-50202-4 Paperback

NOTE BY THE EDITOR

There is given as an Appendix to this volume a complete list, as far as is known, of Sir Charles Parsons' papers from 1885 onwards. From this Collection a fair number of the papers have been omitted in order to avoid reduplication, and for the same reason, amongst others, parts of some of the papers have been left out. It has not been thought necessary to state in each paper where omissions occur, but where it seemed desirable the nature of the matter contained in the deleted part has been briefly indicated. Some repetition has been found unavoidable.

In only a few cases have papers of which Sir Charles Parsons was part author been included in this volume.

CONTENTS

LIST OF PLATES

PREFACE

Sir Charles Parsons, owing to the position which the steam turbine had taken in the world, was much sought after as a speaker or lecturer at public meetings. It cannot be said that he was a good speaker for he was always hampered by a diffidence, amounting almost to shyness, before an audience. But as the author of the greatest mechanical invention of the nineteenth century he was always welcomed and respected, and as he had often something new to reveal he was listened to with attention. Hence, willy-nilly, he was called upon to deliver a great many papers and lectures. At the express wish of Lady Parsons, a selection from these published writings has been made and is here presented.

The first part of the volume is devoted, principally, to papers and lectures dealing with the development and progress of the steam turbine. In making the selection repetition has been avoided as far as possible, and in consequence only a few of the papers are complete. In the second part three lectures on the famous research into the artificial manufacture of diamonds are reprinted in full. They give a complete record of a task which occupied Sir Charles, on and off, for many years.

A third part contains three appendices, of which the last is a list of all Sir Charles' published writings. The other two deal with subjects about which no written records of speeches or addresses remain, but which, nevertheless, it seemed proper to include in this volume. The first describes the Auxetophone, that wonderful air-valve-operated "loud-speaker" on which Sir Charles spent many laborious nights. The monograph was written by the late Mr A. Q. Carnegie to whose devotion, it might almost be said, the saving of the instrument from oblivion is due. This is not the place to speak of Mr Carnegie's attachment to Sir Charles throughout many years and the valuable part he played in the progress of the steam turbine, but it is fitting to express here profound regret that he died before this little tribute to his great master was in type. The second appendix deals with optical glass, the manufacture of which was kept alive in this country largely through Sir Charles' exertions. For the notes upon which it is based thanks must be given to Mr R. S. Campbell, Mr S. M. Morrison and Mr C. Young.

Finally, I desire to thank all the institutions and societies which have courteously permitted extracts to be made from their transactions; my cousin, Mr Arthur Parsons, for the help and advice he has given me throughout; and especially Mr Loughnan St L. Pendred, whose experience in the preparation of text for the press and in the reading of proofs has greatly lightened the work of editing.

But I must not conclude without a final word. Lord Rayleigh has increased the value of this Collection incalculably by a delightful memoir which gives a real picture of Sir Charles as Engineer and Scientist, and also as one who loved the countryside and was the most courteous and kindly of hosts. The memoir was written by Lord Rayleigh at the wish of Lady Parsons, and one cannot but feel that she would have endorsed every word of it had she been spared to see it.

GEOFFRY L. PARSONS

FOREWORD

In offering this collection of papers of the late Honourable Sir Charles Parsons, it is with the hope of helping the interested reader to appreciate the genius, the perseverance, and the indomitable courage of Sir Charles. These were the qualities which carried him through the early days of difficulties and discouragement; through the days of hope and of bitter disappointment, such as when success seemed within grasp only to be foiled by the calamitous wrecks of the *Viper* and the *Cobra*; the bitter grief and sorrow brought him by these disasters; and, finally, to the days, in later years, of brilliant achievement and recognition.

It needed the collaboration of a great scientist to emphasise the distinguishing features in the work of a great engineer. To meet this need Lord Rayleigh generously undertook to write a preface to the papers of his old friend. A preface by so distinguished a scientist will add greatly to the interest of the papers.

Our thanks are also due to the Honourable G. L. Parsons who has undertaken the collecting and sorting of his uncle's papers, a task that entailed nice discrimination and sound judgment.

KATHARINE PARSONS

RAY DEMESNE
KIRKWHELPINGTON
NORTHUMBERLAND

10 *August* 1933

SOME PERSONAL REMINISCENCES OF
SIR CHARLES PARSONS

BY LORD RAYLEIGH

I am proud to have, by Lady Parsons' wish, the opportunity to write a few personal recollections and comments by way of preface to the collected writings of my great and lamented friend, Sir Charles Parsons.

For a connected account of his life, reference will naturally be made to the biography by Mr Rollo Appleyard,* who has admirably worked up the available material. There is, however, very little on record about Parsons' intimate life—in fact, he lacks a Boswell. I am far from pretending to fill the gap; yet, as I was better placed than most, it would be a pity to let slip the opportunity of recording such recollections as remain, notwithstanding that they are somewhat disconnected and only cover a small part of his life and activities. I shall supplement them with recollections by some others of his friends and fellow-workers.

The early struggles of Parsons' career were over at the time when I first got to know him, which was at a garden party at Sir George Darwin's house at Cambridge, when the British Association met there in 1904. He had come through into smooth water, but occasional remarks showed how much he had endured. Thus, apropos of the habit of smoking, he remarked how soothing he had found it at a time of acute financial anxiety. This referred, no doubt, to the litigation over his patents. He appreciated the sympathetic advice of his counsel, Fletcher Moulton,† at that time.

In his address to the Engineering Section of the British Association (1904), he says, "Even in our time I scarcely think that anyone would venture to describe the lot of the inventor as altogether a happy one," and no doubt he was speaking from experience. The young inventor in the commercial world is often regarded as a sort of revolutionary, who wishes to upset vested interests, and many people's hands will at first be against him. Like the political revolutionary, he is apt to be regarded as a dangerous nuisance before he succeeds, and only becomes a hero afterwards.

Parsons' problem had been in many respects one of peculiar difficulty. As Mr Turnbull has well remarked, the turbine was naturally best adapted for large sizes, but had necessarily to be developed through the small sizes, when the reciprocating engine already held the field and was very suitable.

---

* *Charles Parsons*, Constable, 1933.
† Afterwards Lord Moulton of Bank.

There can be little doubt that even among his friends many wise heads were shaken over his determination to pursue its development.

After the Cambridge meeting, I came across C. A. P. on several occasions and the friendship ripened. My first visit to him at Ray, his shooting lodge on the Northumberland Moors, about twenty miles north-west of Newcastle, must have been about 1910 or 1911. I came from the South and he intercepted me at Newcastle, where I had intended to change into the local train, welcomed me kindly, strapped my luggage on to the back of his car, and we motored out. The journey of some twenty miles was not without some psychological interest. Parsons was a severe critic of the drivers of cars that we passed, and called out his criticism in emphatic language. His voice, however, had not much carrying power, and this circumstance may not have been altogether unfortunate. People could not resent what they did not well hear, and if they attempted to reconstruct it for themselves they very likely thought that they had been deceived and that such strong words could not have come from one whose aspect was so mild and benevolent. Sometimes he expressed his protest by blowing his horn, and would continue to blow it for a long time afterwards, much to the surprise of the passers by, who had not seen the earlier phases of the incident.

Ray was at that time as now a house of moderate size and rather lonely situation, but possessing considerable charm and all the essentials of comfort. There was grouse shooting and trout fishing on the loch known as Sweethope, a place of some historic interest as the scene where the Northumbrian rebels of 1715 assembled and pledged their faith to one another. Parsons was fond of both kinds of sport, and he liked the fishing all the better in that he used a motor boat, which frequently refused to work, and gave him the congenial task of dealing with its deficiencies. In the motor boat there was always kept a jar of feathers, and when fishing was done these feathers were used to clean the sparking plug, an operation which was always performed several times during the afternoon.

His sanctum was not without character, and deserves description. There was a writing-table in the window and occasional letters were written there, but not very systematically and for the most part late at night. He never dictated letters. The house-maids were strictly forbidden to touch anything on his writing-table, though Lady Parsons sometimes came in and removed cigarette ends, burnt-out matches, and the like.

At one side of the room was a cage containing a large white cockatoo, and I, for one, must confess to having found its contributions—or rather, interruptions—to the conversation most inharmonious. But Parsons did not seem to be conscious of this. He would get up and placate it with soothing words and with sugar. The cockatoo went to London with him during the winter months.

On the central table was a litter of books and papers—perhaps a number of *Engineering* or *The Engineer*, another of the *Proceedings of the Royal Society*, a copy of the Apocrypha which (oddly enough) was a favourite study with him, and one or two novels, together with pipes and tobacco. *The Pickwick Papers* was a great favourite, and he read it repeatedly. So was P. G. Wodehouse. He liked reading detective stories and "shockers" but I do not think he troubled much to appraise their merits. There was nothing in the way of fixed bookshelves in the study, but a few books were carelessly thrust into a cupboard on one side of the fireplace. Among these were Smiles' *Lives of the Engineers*, some of Rudyard Kipling's works, and a few miscellaneous books on shooting and fishing, and others on scientific subjects.

I never saw anything like an adequate collection of books of reference in any of his houses or in his room at the Works, and I was never able to understand where he turned to for information. It may be suspected that when he wanted it, he simply directed some member of his staff to get it for him. At any rate, he never seemed to suffer from want of being properly informed about any subject which it concerned him to know. He got the necessary knowledge somehow.

To return, however, to the arrangement of his study. On the mantel-piece was a barograph, but the usual printed charts were lacking, and a piece of plain paper on which there were several previous traces would do duty instead. Beside it there lay for many years a unique tobacco pipe, with the bowl made from a stone with a hole in it, which had evidently been picked up. The stem was an ordinary one, introduced into a side hole in the bowl. The temperature in September at Ray, high up on the Northumberland fells, was rarely high enough to be comfortable without a study fire, even in the morning. But Parsons usually lighted it himself, and "drew" it up into a blaze by means of a sheet of newspaper held over the upper part of the opening. He never seemed to be bored with doing this—in fact, he apparently enjoyed it.

In one corner was a gorgeous casket* in which he had been presented with the freedom of the City of Newcastle, and over the mantelpiece was a painting by B. Gribble of the ornate ships of the Spanish Armada. Contrasting with the leisurely dignity of these were photographs of the *Turbinia* cutting her way through the water at full speed, which were hung on the walls. So also were photographs of cavitation produced in the experimental propeller tank at Wallsend, or, perhaps, in the small tank at Holeyn Hall.

If any serious engineering studies were made at Ray, this was not obvious. I never remember seeing a slide rule or a book of logarithms

* This however was later than the date of my first visit.

lying on C. A. P.'s table; and my impression is that he never used either, then or elsewhere, preferring to rely on the multiplication table. Nor, so far as I can recall, were there any engineering drawings or blue prints. Probably, however, he made rough calculations and sketches at home, and took them to the draughtsmen at the Works to be elaborated. On the other hand, mechanical models made of cardboard or paper with corks, cotton reels, knitting needles, wire, sealing wax and string were in evidence, and if a mechanical point were under discussion, he would insist on its being explored and any suggestion being put to the proof, as far as possible in this way. It was much more congenial to his habit of mind to test a question by experiment than by abstract reasoning or calculation.

The practice of making paper models of this kind goes back to very early days. A Cambridge contemporary who wishes to remain anonymous writes:

"He and I were fellow-oarsmen in the Lady Margaret 3rd boat in 1876, and in the first in 1878; it was the custom in those days during the period of training for the members of the crew to breakfast in each other's rooms on alternate mornings, and after breakfast in Parsons' rooms he said, 'Look here, you fellows, I have an engine here which is going to run twenty times faster than any engine to day.' 'Rot!' was the reply; however, on moving towards a side table in the room, there was a 'toy' paper engine into which Parsons blew and the wheels simply flew round, but to show further our contempt we put Parsons and his engine under the table."

On a small table in the study were laid out a few tools of the simple kind which may be bought at the humblest ironmonger's shop. There was no workshop at Ray. At Holeyn Hall, C. A. P.'s earlier home, there was one. I never saw it, but have been told that the lathe and tools were very much neglected, being allowed to get covered with rust from the acid fumes used in his experiments on diamonds.

I can recall many interesting conversations in this room at Ray. We discussed on one occasion the supposed secret possessed by the ancient Egyptians of hardening bronze to a cutting edge. Alan Campbell Swinton, who was present, mentioned a disagreeable legend that bronze swords were hardened by heating them red hot, and holding them against the back of a living slave; but Parsons brought the subject down to earth by remarking that when he spent a winter at Assouan in 1909 he had taken trouble to obtain specimens of ancient bronze chisels, and had found that they had nothing beyond the ordinary degree of hardness that can be attained by cold working.

Once, when he was alone with me and in a confidential mood, he spoke of the disaster of H.M.S. *Cobra*, one of the first destroyers to be fitted with turbine engines, which broke in half in a heavy sea on her voyage south from the Tyne in 1901 under the command of a Naval officer but with some of Parsons' men on board. Parsons was emphatic that the turbine machinery was not in any way the cause of the disaster. His information was, that in the course of the voyage, the deck plates of the ship showed alarming signs of distress and that representations were made to the commanding officer, who declined to listen, as sailors are apt to do when landsmen try to interfere. "If I had been there," said C. A. P., "I would have raised a mutiny and brought her into port"; and his words carried conviction. He was no boaster. Another remark which he made apropos of difficult and dangerous situations should not be suppressed, though it runs counter to the cherished notions of many excellent people. It was that men who were occasionally given to drink had often proved the most helpful to him in emergency. This he connected with a certain recklessness of character which made them willing to take risks in order to save a situation.* "They don't care a damn" he said.

The way in which C. A. P. himself reacted to a difficult situation is well illustrated by a story told by Mr James Denny. "A preliminary trial trip of the *Viper* was made early in the day, and the bearing of the engineers was ominous. They differed with Mr Parsons as to the trial trip rate of wages, and, as the latter knew his own mind, the engineers walked off the ship. Everyone thought that the day's proceedings must end there and then, but Mr Parsons thought otherwise. He turned on his apprentices to do journeymen's work, picked up some men off the quay, borrowed some more from Messrs Hawthorn Leslie and Co., who had the contract for the hull and boilers, and made all into a scratch crew for the trial trip. Under these extraordinary circumstances the *Viper* ran her trial,† and on that day did the unparalleled speed of 37 knots. When Mr Parsons emerged from the engine room, dirty and warm, all crowded round him to congratulate him, but he took the whole thing as a matter of course."

The difficulties which had to be met in the development of high speed propellers were by no means limited to those which were inherent in the problem. Skilled workmen do not like experimental work. The men employed to file up the early high speed propellers felt much disgusted when they were brought back to the shop for alterations. They had a feeling at the back of their minds that their work was being wasted: it was generally agreed that a slow speed propeller was the thing for efficiency. The men may have known or guessed this, and they perhaps thought that Parsons was following a will-o'-the-wisp. Moreover, he was not always

---

* In these controversial matters, I merely repeat the substance of what Parsons said to me, without making any attempt to weigh what might be said against his view.

† I.e. actually the official Admiralty Trial.

reasonably patient with them when they were doing work to which they were not accustomed. Thus, I have it from Mr C. Turnbull, M.I.E.E., that when the rotor for the *Turbinia*, which was the first to have long blades, was under construction, the turner who was turning up the blades took too heavy a cut, and the blades were bent over. When Parsons came into the Works and saw what had happened, he boiled over, and dismissed the unfortunate workman on the spot. This was probably unjust, but it must be remembered that he was naturally temperamental, and that he was living and working at that time under conditions of great strain, financial and otherwise.

This incident brings to mind another of much later date when a workman whom he personally wanted early in the morning was not forthcoming. Parsons became more and more impatient, and when the man ultimately arrived half an hour late he said "You're fired" and would not hear a word in mitigation or excuse. However, a member of the engineering staff insisted on stating the man's case, which was that he had been sitting up all night with a sick wife. Parsons was filled with remorse, decreed that the man's wages were to be increased five shillings a week, and personally called at his home with presents of grapes and other delicacies suitable to an invalid.

I have been tempted to digress somewhat widely from the subject of my own intimate conversations with C. A. P. To return, he once said that the bow of the archer was an admirable contrivance from the point of view of mechanical design. The only part which moved rapidly was the string, which had a negligible mass. Thus little of the energy of the bent bow was wasted in giving velocity to the moving parts of the bow itself, and nearly all went to give velocity to the arrow.

Again, he spoke of his apprenticeship at Elswick and of the first Lord Armstrong, for whom he had a great admiration. Some trouble being experienced with a new device, Parsons asked the foreman of the shop what progress had been made. The answer was "That will soon be all right. He [Armstrong] is attending to that himself." It was not considered prudent, C. A. P. said, to disturb Armstrong when he was grappling with a difficulty. Anyone who was rash enough to do so was likely to regret it.

Mr Appleyard (p. 27) quotes a testimonial received by Parsons from W. G. Armstrong & Co. Nevertheless, it may be doubted whether he was fully appreciated there. His action in taking out a patent was strongly resented by one of the directors, who maintained that he had no right to use material and time belonging to the firm for developing an invention for himself. At that time the doctrine that all such inventions belong to a firm had not definitely crystallised out. The directors were not perhaps predisposed in Parsons' favour by his habit of experimenting with rockets, which sometimes resulted in shattering explosions under the windows of their luncheon room.

I have heard him, too, on the subject of his auxetophone or "bellowphone" as it was colloquially called, used for amplifying the sound of the violin or 'cello. He never made it very clear to me why this invention was dropped. In one mood he seemed to suggest that it was from opposition in the musical world, conceived in a narrow trade-union spirit with which he had no patience. At other times he seemed sympathetic to those who feared loss of employment by the introduction of this instrument, many of whom had written appealing letters to him. It may well be, however, that he was rather tired of negotiating about it, having bigger things on hand.

On another occasion, the Mediaeval Cathedrals regarded as engineering structures were discussed, and someone said that they were an instance of how little practical engineering construction was indebted to formal theory, for they were built before any such theory existed. "I quite agree," said Parsons; "theory is not used in doing the thing. That is put in afterwards to make it look pretty."

It was, in fact, very curious how little use he ever seemed to make of the formal mathematical training he had had at Cambridge. He had been 11th Wrangler in the Mathematical Tripos, and must be credited with having had at one time a thorough mathematical knowledge and facility in using it. At the time I knew him, however, he never allowed this to appear. I do not think that I ever saw him use any mathematical method more complex than the rule of three. If he was invited to listen to anything a little less simple, he always turned it off by saying that his mathematics were rusty. "That is analytical," he would say, "I like something geometrical. I never was very good at analysis and now I have forgotten what I once knew." It was difficult to feel sure how far this could be taken literally, but it is certain that he had no liking for symbolical methods of expression, and seldom, if ever, used them in his published papers. All his turbine calculations, he said, had been made in terms of successive small expansions only, and no use was made of integrated expressions for the work done by the expanding steam. He even affected to regard the use of atomic weights and molecular formulae to determine combining proportions, as a thing beyond his ken.

Later on in his life, he would simply hand a numerical problem to his staff to be dealt with. He took no interest in the mathematical reasoning used by them; he only wanted the result. I never heard him on the subject of entropy, but Lord Falmouth tells me that the mention of it used to fill him with a curious indignation. Apparently he regarded it as a useless mystification. However that may be, he did not object to his staff employing it if they saw fit. Although Parsons did not make use of any formal methods of calculation, he seemed to have a kind of subconscious way of arriving at conclusions for which most people would have found such methods indispensable. Needless to say, he could not explain how he did it—but then he equally failed to explain with any lucidity things which, to the ordinary observer, were much less mysterious.

The Newtonian system of mechanics was, so to speak, part of his religion, and when Einstein and his followers proposed to modify it, Parsons was, I believe, torn by a painful embarrassment. On the one hand, he could not but see that the position of the relativity theory was very strong in that it had quantitatively predicted facts which observation confirmed. When this was pointed out, he did not seriously attempt to dispute it; but he showed that he was not happy in giving up the familiar mental images which he had made his own for others which seemed so vague and elusive. I think it was always a surprise to him that people whom he regarded as much more learned than himself, and who doubtless had academic knowledge which was outside his province, could by no means see with his own sure intuition the way through a mechanical difficulty, and the conditions of success in overcoming it.

The attempts at artificial production of diamonds occupied a large share, perhaps an unduly large share, of Parsons' life work. Contrary to what has sometimes been supposed, he took up this problem from a purely scientific standpoint, and not with a view to manufacturing diamond dust as an abrasive; though, in the event of success, this aspect must ultimately have received attention. His investigations on this subject range over more than thirty-five years, from 1888 onwards. The original idea of melting carbon under the highest temperature and pressure was thoroughly tried out and also the solidification under pressure of carbon dissolved in molten iron, no expense or effort being spared. Dr Stoney estimates that £30,000 was spent on these experiments. Parsons remarked to him, "We have now made a bit of money, and deserve to have some fun." The ultimate results, however, were wholly negative.

Parsons had, in no small degree, the traditional British quality of hating to be beaten, and in this instance it seemed to lead him too far. It was difficult to see much prospect of success in some of his later experiments, in which great trouble and expense were incurred to carry out elaborate furnace operations *in vacuo*. I believe this view was shared by several of his friends, myself among them, but if any of us ventured to press it on him he would only reply, "I think it ought to be tried." It was not perhaps surprising that he should rely on his own judgment rather than on the judgment of others. In his youth the prospect of constructing an economic rotatory engine was generally considered visionary. Innumerable inventors had attempted it from the time of James Watt onwards; but in the face of all scepticism Parsons had shown the engineering world that he could succeed where others had failed. Parsons' persistence in the research on diamonds was no doubt encouraged by the reported positive results of Moissan and Crookes, which had long been generally accepted in this country. After Crookes' death some of his specimens came into Parsons' hands; he tested them, and came to the conclusion that they would not burn in oxygen and therefore could not be diamonds. So far as I am aware, this result has never been made public. He also explored the ground in Paris and tried to get on the track of Moissan's specimens: but he could not learn that any such had been preserved. He found moreover that some of Moissan's old colleagues did not feel any confidence in his results, though (one may suppose) they had for a time naturally shown some reticence on the subject.

During my visits to Ray, I often went with Parsons to his Works at Heaton and at Wallsend, and lunched with him and his staff at the former place. So far as I saw on the occasions of my visits, he did not work much in his own room there. This had a rather neglected aspect, and indeed, though he must have roughly sketched specifications of patents and written out the papers that were published by the various engineering and scientific societies, I never saw him doing it, nor got any insight into his methods of composition. He showed me the manuscript of his Presidential Address to the British Association (whether typed or not I do not remember) and asked for criticism. I pencilled what seemed to me slight verbal improvements in a few passages. These he professed to adopt; but I was amused to observe that when the address came out, the suggested alterations had been deleted and the original wording restored. This was an example of a trait in him that others have noticed. Lord Moulton, who had the greatest admiration for Parsons, remarked that he might seem to be impressed by what was said to him, but, often enough, quietly retained his original opinion.

Lady Parsons remembered once finding him in his own room at the Works reading the newspaper. This struck her as so unusual that she was moved to wonder what was wrong. It appeared that things had come to his notice which forced him to conclude that the organisation of the Works was seriously amiss, and he was painfully perplexed as to what to do about it. Reading the paper, one may suppose, served as a sort of mental anodyne, while the problem was being battled with subconsciously.

He once remarked to me—apropos not of any one in particular, but of the selection of his right-hand men generally—that it was hard to find anyone under 50 years old who was much good. It is fair to observe, however, that he was over 70 himself at the time. At an earlier date he would perhaps have put the limit rather lower.

Parsons, like most inventors, was alive to the importance of secrecy in matters which were not protected by patents. In the early days, when he was interested in the manufacture of carbon incandescent lamps, a carbonaceous mixture containing sugar was in use for making the squirted filaments, and, for fear someone should taste it, and learn the secret, the bottle was labelled ARSENIC. The same spirit was observable at Heaton Works. Thus, when C. A. P. had shown me a wooden pattern of his latest design of propeller and explained his notions about it, he said, "It will not do to leave it about where everyone can see it," and, so saying, he hid it away behind the books in the bookcase.

Again, at my first visit, I was given to understand that no one was ever admitted into the department where searchlight mirrors were made, and when I inadvertently got too near the subject in conversation, I was gently warned off. At a later visit, this embargo was removed, and I was offered the opportunity of seeing everything. I gathered from the staff and from Parsons' manner that this was a high favour. The main secret has long since leaked out, that the mirrors were made from plate glass by softening it at a moderate heat and shaping it to a paraboloidal metallic former. It was then carefully annealed. The mirror thus made was ground and polished on a machine by a comparatively small tool, the surface of the glass itself serving as a guide. The general shape given in the moulding process was not seriously interfered with in the grinding and polishing. Parsons had recently examined the annealing process, and had determined by the permanent bending of a glass thread what was the temperature at which it became plastic. This investigation pleased him very much by its simplicity and showed the way to a considerable economy of fuel in the annealing process. To amuse visitors he liked to bring one of the largest (60-inch) mirrors out into the sunlight, and ignite a large piece of wood at the focus.

Parsons was a singularly modest man, and did not seem at all to realise his own standing in the world. I believe he appreciated the distinctions which were conferred upon him but did not draw the obvious inference, or at any rate did not keep it in mind. His want of self-assertion was at times almost comic. Thus, when he went to the International Astronomical Union at Rome in 1922, he failed to receive the customary invitations to the various receptions. He seemed to think that this was a sign that he was not appreciated, and was much hurt. He did not think proper to take any action, and returned home only to find the missing invitations awaiting him there! Again, he has been known to listen to a ship's engineer explaining the action of a steam turbine without giving the slightest hint that he had ever heard of such a thing before! In some people, silence under such circumstances might have been taken as rather disagreeable and sardonic. In his case, it was simply an indication of a very retiring disposition. Nevertheless he was not always prepared to be silent when he did not agree. On one occasion he fell into conversation at the Northern Counties Club, at Newcastle, with a naval officer, the only other diner. "Who is that contradictious old gentleman?" the latter asked afterwards. "He seems to think he knows more about the engines of my ship than I do myself." It is a safe guess that he did know very much more.

C. A. P. was at times absent-minded in the affairs of everyday life. This was attributed by his friends and neighbours to his being absorbed in thought. In matter of dress he was as a rule fairly tidy, but with occasional lapses. Thus, invited to meet his Sovereign at a great house, he created some surprise by coming down to dinner without a tie: and other incidents of a similar kind could be recalled, if it were thought worth while. But one of the most amusing instances of his absent-mindedness was shown at the house of his neighbour Mr Joseph Straker. It was the custom of the house to collect scraps from the breakfast table, such as fragments of bacon, bones, leavings of bread, porridge, and the like, in a pie dish, and to cover it with milk and gravy for the dogs. C. A. P. came down late, after most of the other members of the party had finished, and the customary collection of scraps had been made. He was observed to go to the side table, help himself from the dog's dish, eat his way steadily through what he had taken and come back for more!

He had a strong dislike of adopting any contrivance from foreign engineering design. His staff were sometimes compelled to adopt unsatisfactory alternative expedients in order to fall in with this sentiment. He would be told that a foreign firm met the difficulty in such a way. "Very well, then," he would say, "let's do something different." This instinctive dislike of conventionality in the way of doing things was often to be seen. Lady Parsons remembered that their daughter, Miss Rachel Parsons, when about 15 years old brought home a mathematical problem which she had solved, but for which marks had been denied her, because the solution was not considered orthodox. C. A. P. followed her solution and approved it. He went on to complain that school teachers were always "down" on any originality of thought, and discouraged their pupils from wasting time on what they called "useless" experiments.

During the war I frequently sat with Parsons on committees at the Admiralty Board of Invention and Research and elsewhere. His want of readiness and lucidity of speech hampered his usefulness in this work, and often practically prevented his views from carrying their proper weight. He was painfully conscious of this, and was mortified by it. It was difficult or impossible to persuade him that others, who were more vocal and made their views heard when his were practically a sealed book, did not mean him any injury. But these resentments did not last long and vanished like a puff of smoke if a friendly advance came from the other side.

Throughout his life it was very noticeable how much pleasure he took in "tinkering" and doing small mechanical repairs with his own hands. As the motor car developed from its early beginnings, his pleasure in using it seemed to diminish in the same ratio as the number of mechanical breakdowns which he was called on to deal with by the roadside. He was ready with expedients which to the ordinary motorist would seem fantastic. A fire was on occasion lit at the roadside, and a bent axle heated red hot, and hammered straight with a stone. Once in bitterly cold weather, the car refused to start, and no ordinary means were at hand for warming it. C. A. P. reflected, however, that man's animal economy provides periodical supplies of a warm fluid: and this source of heat was applied to the outside of the carburetter with good effect.

Again the Parsons family were to rendezvous with Mrs J. H. Cuthbert's* party for a picnic. They arrived behind time. Parsons' explanation to her was somewhat cryptic. He said: "The car would not go, and we had to undress the parlour maid." It appeared that a piece of steel spring was required to repair the distributor, and some mysterious intuition told C. A. P. that the fastening of the parlour maid's stays would afford what he wanted! She was accordingly sent upstairs to undress, and the stays were sacrificed. He often wanted to requisition Lady Parsons' engagement ring when he was at work on the diamond problem: but I believe that in this case his wishes did not prevail so easily.

At Ray, the house was lighted by a small direct-driven petrol motor set without accumulators. At times it gave trouble, and Parsons would get up from the dining-table in his dress clothes and go out to deal with it himself. One felt that an elderly man who would do this rather than order in candles was indeed a born mechanic. A machine which would not work seemed to be the thing of all others that stimulated him into activity. When he was no longer young, his friends would view with concern and disapprobation the energy he could put into "cranking up" a refractory motor car. He never seemed to take the slightest interest in, or care for, his own health; though if anyone about him either at home or at the Works was unwell, he was most solicitous that they should have the best advice.

He had a boyish pleasure in any experiment which ended in a big bang or flare up, such as firing a rifle bullet into a block of steel a few inches from the muzzle. He made many experiments of this kind in connection with the diamond problem. Once when something broke down

* Now Lady Rayleigh.

under extreme experimental conditions, I remember the characteristic words, "Well, we have the satisfaction of having bust it."

Going back to the time before I knew him, and when his own children and those of his friend and neighbour Mr Norman Cookson were in their 'teens, he was seen in his best and most genial mood when playing with them. He constructed model machines of original design for their amusement, and, I suspect, incidentally for his own. The helicopter illustrated in Mr Appleyard's book (p. 66) was one of these; and if the children mishandled or broke them it never ruffled him, any asperities in his character being for the time completely in abeyance. They were encouraged to come into his workshop at Holeyn and "help" him as they called it, when he was working at experiments on cavitation, the auxetophone, and so on.

There was a small trout stream close by, and he delighted in going there with them provided with spades and buckets. A bend of the stream would be dammed up; and short-circuited by a canal cut across. The trout would thus be stranded and captured. This, if not sportsmanlike in the most orthodox sense, was, to the children, the greatest fun that could be imagined.

It would be a mistake, however, to think that he did not appreciate orthodox sport. In his early days he appeared with the Tynedale hunt in surprisingly correct get-up—pink coat, top boots, and so on. Probably Lady Parsons' hand may be traced in this. He showed no lack of pluck in riding to hounds, though his horsemanship might be rudimentary. He was fond of entertaining his friends and neighbours for the shooting at Ray, and their enjoyment was his chief concern. A more unselfish host could not be imagined.

He liked occasional shooting as well or better than regular grouse drives. This recalls an amusing incident which I have from Lady Rayleigh. He went for a picnic to Sweethope on one occasion with a party of young people, and took a gun with him. A grouse was shot on the way across the heather. When they had arrived by the lake, and were about to lunch, someone said "I suppose if we were gipsies we should cook the grouse in clay and eat it, like they do the hedgehogs." The idea appealed to C. A. P. at once. A fire was lighted, and clay was ruthlessly dug out with a silver spoon, the only tool which was to hand. The grouse, feathers and all, was covered with a layer of butter, and an outer coating of clay, and put on to bake. After a due interval, the clay ball was opened and the grouse came away from the feathers. He ate his share of it with much contentment.

Parsons kept up his interest and enthusiasm to the end, though it seems that he was somewhat depressed to find that, as was inevitable in so large a concern, many of the problems at the Works had passed beyond his personal guidance and control. He was full of schemes for the improvement of large telescope mirrors, and, within a day or two of his death, had dwelt on his hopes of "having a shot" at making the mirror of 200-inch diameter required by the American Committee. He had evidently no notion that the end was so near.

I bring these random recollections to an end with some regret that it has not proved possible to weave them into a more systematic narrative. It has been unavoidable to jump from one scene to another, with little continuity of time, place or subject. The purpose has been to bring before readers of this volume the personality of one of the greatest figures in the engineering world of his own or any other time. To read these recollections cannot, however, confer the great privilege which was enjoyed by those who knew him.

PART I

THE COMPOUND STEAM TURBINE AND ITS THEORY, AS APPLIED TO THE WORKING OF DYNAMO-ELECTRIC MACHINES

North-East Coast Institution, December 19th, 1887

[The paper opens with a short discussion of the principles of water turbines. That part is omitted here.]

The essential conditions for high efficiency are, as we have said, absence of shock and low residual velocity. These conditions, in the case of the water turbine, can be better obtained in the types called the outward or inward flow than in the parallel flow, for the large volume of water to be dealt with necessitates long blades of double curvature for the last-mentioned type, these are objectionable, while it is more difficult to minimise the residual velocity of the water leaving the wheel. Principally these reasons have led to the more general adoption of the outward or inward flow types than of the parallel flow. After careful consideration, however, the parallel flow type seemed more suitable to the compound steam turbine, and has been accordingly adopted.

Fig. 2 shows the arrangement of ninety complete turbines, forty-five lying on each side of the central steam inlet. The guide blades R are cut on the internal periphery of brass rings, which are afterwards cut in halves and held in the top and bottom halves of the cylinder by feathers; the moving blades S are cut on the periphery of brass rings, which are afterwards threaded and feathered on the steel shaft, and retained there by the end rings which form nuts screwed on to the spindle. The whole of this spindle with its rings rotate together in bearings, shown in enlarged section (Fig. 3). Steam entering at the pipe O flows all round the spindle and passes along right and left, first through the guide blades R, by which it is thrown on to the moving blades S, then back on to the next guide blades, and so on through the whole series on each hand, and escapes by the passages P at each end of the cylinder connected to the exhaust pipe at the back of the cylinder. The bearings (Fig. 3) consist of a brass bush, on which is threaded an arrangement of washers, each successive washer alternately fitting on to the bush, and the block, while being alternately 1/32nd smaller than the block outside, and 1/32nd larger than the bush in the hole. One broad washer at the end holds the bearings central. These washers are pressed together by the spiral spring N and nut, and by friction against each other, steady or damp any vibration in the spindle that may be set up by want of balance or other cause at the high rate of speed that is necessary for economical working.

The bearings are oiled by a small screw propeller I attached to the shaft. The oil in the drain pipes D and F, and the oil tank D lies at a lower level than the screw, but the suction of the fan K raises it up into the stand pipe H over and around the screw, which grips it and circulates it along the

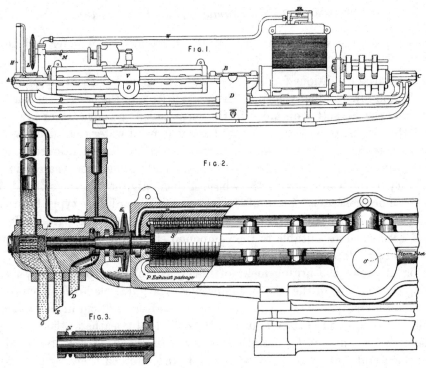

Figs. 1, 2 and 3

pipes to the bearings. The course of the oil is as follows: The oil is forced by the propeller I and oils the bearing A, the greater part passes along the pipe E to the end bearing C, some, after oiling the bearing C, drains back by the pipe F to the reservoir D, the remaining oil passes along the armature spindle, oils the bearings B, and drains into the reservoir D, from which the oil is again drawn along the pipe G into the stand pipe H by the suction of the fan K. The suction of the fan is also connected to the diaphragm L and forms, with it and the spring M the principal part of the governor which actuates the throttle valve V. Fig. 4 is the electrical control governor, which will be further described in connection with the dynamo. It acts directly upon the controlling diaphragm L by admitting or closing a large access of air to it, and thus exercises a controlling influence upon it.

For small differences of pressure the velocity of air or steam passing orifices or short tubes is given by the same formula as that for water. The heads of pressure have to be calculated in each case in terms of the respective fluids, and for small differences of pressure the velocity of efflux is the same for the same head in all cases. This has been recognised as a fundamental principle; it has also been verified in some of the cases we are going to consider by actual measurements.

We now proceed to consider the action and condition of the steam as it passes through the successive turbines, but, before doing so, it will be necessary to go into some calculation in order to form a correct basis to start from.

The velocity of efflux of steam flowing from a vessel at 15·6 lb. per square inch absolute pressure through an orifice into one at 15 lb. absolute pressure is 366 feet per second, the drop of pressure of 0·6 lb. corresponding to a diminution of volume of 4 per cent. in the opposite direction. We must now suppose that the turbines are so proportioned and the blades so diminished in size that each one nearer the steam inlet, O, has 4 per cent. less blade area or capacity than the preceding one all the way from each of the exhaust ends up to the centre.

If we work this out we find that after forty-five turbines the pressure will be about 69 lb. above the atmosphere of the inlet. The steam will have followed some curve of temperature and volume very near the adiabatic but on the isothermal side of it; the velocity of flow will be slightly greater near the inlet in consequence of the increased temperature. The steam enters from the steam pipe at 69 lb. pressure and passes through the first turbine, consisting of a ring of guide blades and then a ring of moving blades, and falls 2·65 lb. in pressure; the velocity due to this fall in pressure is 386 feet per second; its volume increases by 3·85 per cent. of its original volume; it passes the second turbine, falls 2·55 lb., and again increases its volume in the same ratio, and so on till it reaches the last turbine, where its pressure is 15·6 lb. before entering and 15·0 lb. on leaving it to flow into the exhaust pipe. The velocity due to the last drop is 366 feet per second. The steam has therefore nearly the same velocity of flow for all the turbines; in other words, the error in assuming the velocity due to the head to be 376 feet per second throughout does not exceed 3 per cent. The velocity of the wheels at 9200 revolutions per minute is 150 feet per second, or 39·9 per cent., of the mean velocity due to the head throughout the turbines.

On comparing this ratio of velocity of wheel to velocity of flow with that of the Tremont water turbine we find a corresponding efficiency of a little over 72 per cent. Hence in the compound turbine we are describing, the velocity of the blades is sufficient to secure a very high return of useful effect. We may, therefore, assume that if the blades be equally well shaped in the steam turbine as in the water turbine, and that the clearances be kept small and the steam be dry, then each turbine of the set will give an efficiency of at least 72·5 per cent. We say "at least" because as each turbine discharges without check into the next, the residual energy after leaving the moving blades is not lost as it is in the case of the water turbine we are considering (it amounts to from 3 to 5 per cent. of the energy), but continues into the next guide blades, and is wholly utilised there in assisting the flow. This unchecked flow from one turbine to another is a great advantage of the parallel flow type over other types where a number of turbines are compounded. It also minimises the skin friction, as nearly the whole of the moving part is covered with blades, and there is consequently no appreciable loss due to this cause.

As each turbine of the set gives 72·5 per cent. efficiency it follows that, as the steam at all points expands gradually without shock, the motor as a whole will give an efficiency of at least 72 per cent. of the total mechanical energy of the steam, or in other words, over 72 per cent. of the power derived from using the steam in a perfect engine, without losses due to condensation, clearances, or friction, and such like. A perfect engine working with 90 lb. boiler pressure, and exhausting into the atmosphere, would consume 20·5 lb. of steam per hour for each horse-power; a motor giving 70 per cent. efficiency would therefore require 29·29 lb. of steam per horse-power per hour.

We have now gone sufficiently into the question to show clearly that in the compound steam turbine we have a motor whose theoretical efficiency is very high indeed. The fact that at each turbine of the set the temperature is constant, enables us to predict that there can only be a very minute loss arising from condensation. This has been conclusively verified by experiments with superheated steam, coupled also with the fact that "for small differences of pressure gases and vapours act like liquids in flowing through orifices and tubes, in virtue of the small differences of pressure, and that the velocity of flow is regulated by the fundamental formula $v = 8 \sqrt{n}$". These three facts draw the analogy between the water turbine and the compound steam turbine for efficient working as close as it is possible to draw anything, and it does not seem too much to expect, in the larger sizes at any rate, an equally good efficiency.

The dynamo which forms the other portion of the electric generator (Fig. 1) is coupled to the motor spindle by a square tube coupling fitted on to the square spindle ends. The armature is of the drum type, the body is built up of thin iron discs threaded on to the spindle and insulated from each other by tracing-paper. This iron body is turned up and grooves milled out to receive the conducting wires. For pressures of 60 to 80 volts there are fifteen convolutions of wire, or thirty grooves.

The wire starting at *b* is led a quarter of a turn spirally, *c*, round the cylindrical portion *a*, then passing along a groove longitudinally is again led a quarter turn spirally, *d*, round the cylindrical portion *a*, then through the end washer and back similarly a quarter turn over *e*, then led along the diametrically opposite groove, and lastly, a little over a quarter turn *f*, back to *g*, where it is coupled to the next convolution (Fig. 5).

The commutator is formed of rings of sections; each section is formed of short lengths; each length is dovetailed and interlocked between conical steel rings; the whole is insulated with asbestos, and when screwed up by the end nut forms with the steel bush a compact whole. There are fifteen

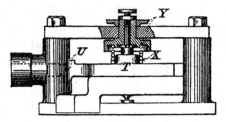

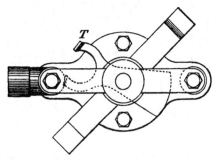

Fig. 4

sections in the commutator, and each coupling is connected to a section. The whole armature is bound externally from end to end with brass or pianoforte steel wire. The magnets are of soft cast iron and of the horseshoe type; they are shunt wound only.

On the top of the magnet yoke is the electrical control governor (Fig. 4). It consists of one moving spindle on which are keyed a small soft iron bar, and also a double finger *T*. There is also a spiral spring *X* attached at one end to the spindle, and at the other to an adjustable top head and clamping nut *Y*.

The double finger *T* covers or opens a small hole in the face *U*, communicating by the pipe *W* to the diaphragm *L*.

The action of the magnet yoke is to attract the needle towards the poles of that magnet, while, by turning the head, the spiral spring X is brought into tension to resist and balance this force, and can be set and adjusted to any degree of tension. The double finger T turns with the needle, and by more or less covering the small air-inlet hole U, it regulates the access of air to the regulating diaphragm L. The second finger is for safety in case the brushes get thrown off, or the magnet circuit be broken, in which case the machine would otherwise gain a considerable increase of speed before the diaphragm would act. In these cases, however, the needle ceases to be attracted, falls back, and the safety finger closes the air-inlet hole.

There is no resistance to the free movement of this regulator. A fraction of a volt increase or decrease of potential produces a considerable movement of the finger, sufficient to govern the steam pressure, and in ordinary work it is found possible to maintain the potential within one volt of the

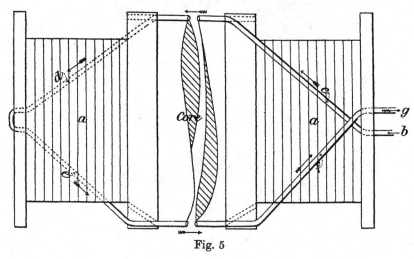

Fig. 5

standard at all loads within the capacity of the machine, excepting only a slight momentary variation when a large portion of the load is switched on or off.

The resistance from brush to brush is only 0·0032 ohm, the resistance of the field magnets is 17·7 ohms, while the normal output of the dynamo is 200 amperes at 80 volts. This, excluding other losses, gives an efficiency of 97 per cent. The other losses are due to eddy currents throughout the armature, magnetic retardation, and bearing friction. They have been carefully measured.

By separately exciting the field magnets from another dynamo, and observing the increased steam pressure required to maintain the speed constant, the corresponding power was afterwards calculated in watts.

The commercial efficiency of this dynamo, after allowing for all losses, is a little over 90 per cent. In the larger sizes it rises to 94 per cent.

Assuming the compound steam turbine to give a return of 70 per cent. of the total mechanical energy of the steam, and the dynamo to convert 90 per cent. of this into electrical output, this gives a resulting efficiency of 63 per cent.

As steam of 90 lb. pressure above the atmosphere will with a perfect non-condensing engine give a horse-power for every 20·5 lb. of steam consumed per hour, it follows that an electrical generator of 63 per cent. efficiency will consume 32·5 lb. of steam for every electrical horse-power per hour.

Again, with steam at 150 lb. pressure above the atmosphere a generator of the same efficiency would consume only 22·2 lb. of steam per electrical horse-power per hour.

The results we have so far actually obtained are a consumption of 52 lb. per hour of steam for each electrical horse-power with a steam pressure of 90 lb. above the atmosphere. We, however, expect shortly to obtain results more nearly coinciding with those very high economies that theory has led us to believe possible with this system. In the larger sizes, so far as we have yet gone, we have invariably found increased economy, as in them the clearances are proportionately less and it is easier to arrange the distribution of steam.

Exceptionally low steam consumption is not the only essential attribute of an electric generator, there are other considerations, in most cases of more importance; these are steadiness of the current produced, freedom from accident, and simplicity, small first cost and cost of upkeep, little attention required, smallness of size and weight for a given output, and an almost insignificant consumption of oil.

To illustrate this we will take the installation at the Phoenix Mills, Newcastle-on-Tyne, which has been running on the average 11 hours daily for the last two years.

Out of the original 159 Edison Swan 16-candle-power lamps 65 are still in good condition, having run 6500 hours at about standard brightness. Now, if the lamps had only lasted 1000 hours on the average, the cost of lamps renewed would have amounted to about double the year's cost of fuel as at present consumed.

At the Newcastle-on-Tyne Industrial Exhibition thirteen of these turbo-electric generators lighted the whole of the courts, giving a total of about 280 electrical horse-power.

During the whole run of the Exhibition the only noticeable accident which occurred to the installation was due to the blowing out of the packing in a branch steam pipe connecting another engine to the main steam pipe supplying the generators, but as there was no stop valve in this branch pipe, the engineer in charge deemed it prudent to shut off steam at the boilers, thus stopping all the generators lighting the four courts for about three-quarters of an hour; this, however, was an accident quite external to the generator and, as it happened, external to the main steam pipe supplying them.

We believe that there were only three occasions during the run of the Exhibition when a small group of lamps was extinguished for the few seconds necessary to enable a spare generator to be started and switched on to them.

The seventeen generators were in two groups; had they been placed in one group the results showed that on all occasions one spare generator would have been sufficient to keep every lamp alight.

In regard to simplicity and cost of upkeep, we may say that in a 30 horse-power generator there is no part that three men cannot lift, and that every part is immediately accessible.

After two years' working, of 10 hours daily, we have found the wear very small indeed, in some cases almost inappreciable, the blades, or vanes, show no signs whatever of any cutting action of the steam.

The commutators in the larger generators have generally withstood this amount of wear, while in some cases where they have been carefully attended they have suffered very little wear indeed.

We have at the present time some important experiments being carried out in regard to the generators. It would, however, now be premature to foreshadow the probable results of these trials. We may, however, say that they lead us to conclude that, most especially in the larger sizes, very much increased economy will be obtained, amounting to a reduction of from 20 to 25 per cent. of the steam consumption quoted in the paper.

We may also add that we have now supplied generators giving a total output of 1280 electrical horse-power; about two-thirds of this amount has been for ships, and one-third for land installation. This amount does not include the Newcastle Exhibition.

We have supplied generators to the English, Austrian, Italian, Spanish, Chilian, Chinese, and United States Governments. Figs. 1 and 2 are views of a typical machine showing its construction. Fig. 3 is the old flexible bearing with washers. Fig. 4 is the governor on the top of the magnets. Fig. 5 is the drum winding of the armature.

In the course of his replies to the discussion Sir Charles said: "In regard to the arrangement of turbines for high pressures, such as 150 lb. per square inch, he would say that in the larger sized generators they had adopted an arrangement called the triple compound type. In this type the turbines or wheels increased in diameter by steps towards the exhaust ends. By this means the proper distribution of steam at very high pressures could be easily dealt with in a satisfactory manner".

THE APPLICATION OF THE COMPOUND STEAM TURBINE TO THE PURPOSE OF MARINE PROPULSION

Institution of Naval Architects, April 8th, 1897

It has been suggested by Sir W. H. White that a paper giving some account of the application of the compound steam turbine to the purpose of marine propulsion might be of interest to the members of this Institution.

The date of this paper is, perhaps, somewhat premature, as the *Turbinia*, the first boat fitted with turbine engines, has not as yet completed her experimental trials, but as the results so far ascertained are in some respects remarkable, this, perhaps, may afford some excuse for their publication.

The manufacture of the compound steam turbine was first commenced in the year 1885, with the construction of small engines for the driving of dynamos; successive improvements were made, and larger engines constructed, but up to the year 1892 the consumption of steam was not such as to justify the application of this class of engine to the purpose of marine propulsion, though, on account of its light weight, small size, and high speed of revolution, it presented great advantages over ordinary engines for certain classes of work.

In the year 1892, however, a highly developed compound turbine, adapted for condensing, was constructed for the Cambridge Electric Supply Company, and when tested by Professor Ewing, F.R.S., showed a consumption of steam equivalent to 15·1 lb. per indicated horse-power per hour, the boiler pressure being 100 lb., and the steam superheated to 127° F. above the point of saturation.

More recently compound turbine engines have been constructed up to 900 horse-power, both condensing and non-condensing, and consumptions of steam as low as 14 lb. per indicated horse-power with saturated steam, and 100 lb. boiler pressure, have been ascertained in engines of 200 horse-power, and still lower consumptions in engines of larger size. Many of the original engines are still doing good work; some, especially the larger sizes of 500 horse-power and upwards, are frequently kept at work for several weeks without stopping. The returns of the Newcastle and District Electric Lighting Company show a yearly cost of upkeep of 2½ per cent. per annum, and the total horse-power of turbines now at work in England exceeds 30,000 horse-power.

In January, 1894, a syndicate was formed to test thoroughly the application of the compound steam turbine to marine propulsion, and a boat was designed for this purpose. In view of the large amount of alteration that would probably be required before a satisfactory issue was reached, and the large amount of time and expense necessarily involved, it was decided to keep the dimensions as small as possible, but not so small as to preclude the possibility of reaching an unprecedented rate of speed, should all the parts work as satisfactorily as was anticipated.

The fulfilment of these anticipations was, however, much delayed, and almost frustrated, by a difficulty which, though foreseen, proved to be of a much more serious character than was anticipated. This difficulty was that termed by Mr R. E. Froude "the cavitation of the water", or, in other words, the hollowing out of vacuous spaces by the blade of the screw, and this pitfall for the designers of screws for very fast vessels, though indicated by theory to exist, came upon us in the case of our very fast-running screw, taxed beyond the usual extent, in its most aggravated form. When the boat and machinery were designed, the trials of the *Daring*, which first drew attention to this difficulty, had not taken place.

The *Turbinia*—as the boat is named—is 100 feet in length, 9 feet beam, and $44\frac{1}{2}$ tons displacement. The original turbine engine fitted in her was designed to develop upwards of 1500 actual horse-power at a speed of 2500 revolutions per minute. The boiler is of the water-tube type for 225 lb. per square inch working pressure, with large steam space, and large return water legs, and with a total heating surface of 1100 square feet, and a grate surface of 42 square feet; two firing doors are provided, one at each end. The stokeholds are closed, and the draught furnished by a fan coupled directly to the engine shaft. The condenser is of large size, having 4200 square feet of cooling surface; the circulating water is fed by scoops, which are hinged and reversible, so that a complete reversal of the flow of water can be obtained should the tubes become choked. The auxiliary machinery consists of main air pump and spare air pump, auxiliary circulating pump, main and spare feed pumps, main and spare oil pumps, also the usual bilge ejectors; the fresh-water tank and hotwell contain about 250 gallons. (See Plate III.)

The hull is built of steel plate, of thickness varying from $\frac{3}{16}$ inch in the bottom to $\frac{1}{16}$ inch in the sides near the stern, and is divided into five spaces by water-tight bulkheads.

The approximate weights are:

		Tons
Main engines ... ...3 tons 13 cwt.		
Total weight of machinery and boiler, screws and shafting, tanks, etc.		22
Weight of hull complete		15
Coal and water...		$7\frac{1}{2}$
	Total displacement	$44\frac{1}{2}$

Trials were made with screws of various patterns, but the results were unsatisfactory, and it was apparent that a great loss of power was taking place in the screw.

To investigate the question thoroughly, a spring torsional dynamometer was constructed, and fitted between the engine and screw shaft, measuring the actual torque transmitted. The measurements conclusively proved that the cause of failure lay entirely in the screws, and, with the object of further investigating the character of this waste of power, a series of experiments was made with model two-bladed screws of 2 inches diameter, revolved in a bath of water heated to within a few degrees of the boiling point, and, in order that the model screw should produce analogous results to the real screw, it was arranged that the temperature of the water and the head of water above the propeller, as well as the speed of revolution, should be such as to closely resemble the actual conditions and forces at work in the real screw, the object in heating the water being to obtain an increased vapour pressure from the water, so as to permit a representation of the conditions with a more moderate and convenient speed of revolution than would otherwise have been necessary.

The screw was illuminated by light from an arc lamp reflected from a revolving mirror attached to the screw shaft, which fell on it at one point only of the revolution, and by this means the shape, form, and growth of the cavities could be clearly seen and traced as if stationary. It appeared that a cavity or blister first formed a little behind the leading edge, and near the tip of the blade; then, as the speed of revolution was increased, it enlarged in all directions until, at a speed corresponding to that in the *Turbinia's* propeller, it had grown so as to cover a sector of the screw disc of 90°. When the speed was still further increased, the screw, as a whole, revolved in a cylindrical cavity, from one end of which the blades scraped off layers of solid water, delivering them on to the other. In this extreme case nearly the whole energy of the screw was expended in maintaining this vacuous space. It also appeared that when the cavity had grown to be a little larger than the width of the blade, the leading edge acted like a wedge, the forward side of the edge giving negative thrust.

From these experiments it would appear that in all screws, of whatever slip ratio, there will be a limiting speed of blade, depending upon the slip ratio and the curvature of the back—in other words, on the slip ratio and thickness of blade; beyond this speed a great loss of power will occur; and that, should the speed of ships be still further increased, the adoption of somewhat larger pitch ratios than those at present usual will be found desirable. It is not proposed here to trace further the losses of power by cavitation, but, generally speaking, the effect is felt in the case of the real ship, not in the racing of the screw, but in loss of propulsion effect.

In the model experiments, however, in hot water, the effect was both loss of propulsion effect and also racing, as would naturally be expected from the fact of greater vapour density of the water in the latter case rendering the cavities more stable.

A series of model experiments on cavitation in cold water on the lines described would be extremely interesting, and probably instructive, but would require more elaborate, powerful, and extremely high speed apparatus than was at our disposal. It would also seem that the limitation imposed on slip ratio tends in favour of larger pitch ratio for very fast vessels.

The single compound turbine engine was now removed from the boat and replaced by three separate compound turbines, directly coupled to three screw shafts, working in series on the steam, the turbines being the high-pressure, intermediate, and low-pressure, and designed for a complete expansion of the steam of 100-fold, each turbine exerting approximately one-third of the whole power developed, the three new screw shafts being of reduced scantling. By this change the power delivered to each screw shaft was reduced to one-third, while the division of the engine into three was favourable to the compactness and efficient working of the turbines. The total weight of engines and the speed of revolution remained the same as before. The effect on the screws was to reduce their scantling, and to bring their conditions of working closer to those of ordinary practice. The thrust of the propellers is balanced by steam pressure in the motors. The rest of the machinery remains the same, though some changes in arrangement were necessary. The usual lignum-vitæ bearings are used for the screw shafts. The engine cylinders lie closely to the bottom of the boat, and are bolted directly to small seatings on the frames of sufficient strength to take the thrust of the propellers. The centre of gravity of the machinery is consequently much lower than with ordinary engines.

At all speeds the boat travels with an almost complete absence of vibration, and the steady flow of steam to the motors may have some influence on priming; at any rate, no sign of this has yet occurred with ordinary Newcastle town water. No distilling apparatus has been fitted. The boat has been run at nearly full speed in rough water, and no evidence of gyroscopic action has been observable.

The oiling of the main engines is carried on automatically under a pressure of 10 lb. per square inch by a small pump worked off the air-pump engine; a small independent duplex oil pump is also fitted as standby. The main engines require practically no attendance beyond the regulation of a small amount of live steam to pack the glands and keep the vacuum good.

The advantages claimed for the compound steam turbine over ordinary engines may be summarised as follows:

1. Increased speed.
2. Increased economy of steam.
3. Increased carrying power of vessel.
4. Increased facilities for navigating shallow waters.
5. Increased stability of vessel.
6. Increased safety to machinery for war purposes.
7. Reduced weight of machinery.
8. Reduced space occupied by machinery.
9. Reduced initial cost.
10. Reduced cost of attendance on machinery.
11. Diminished cost of upkeep of machinery.
12. Largely reduced vibration.
13. Reduced size and weight of screw propellers and shafting.

APPENDIX

TRIALS OF THE *TURBINIA*

In December of last year several runs were made on the measured mile, and the maximum mean speed obtained after due allowance for tide was 29·6 knots, the mean revolutions of the engines being 2550 per minute. Since then new propellers of increased pitch ratio have been fitted.

Further trials were made on April 1st. The mean of the two consecutive runs gave a speed of 31·01 knots, the mean revolutions of the engines being 2100 per minute, the fastest run being at the rate of 32·61 knots.

The utmost horse-power required to drive the boat at the speed of 31·01 knots is 946, as calculated from experiments on her model, made at Heaton Works, on the method of the late Mr William Froude. Assuming the ratio of thrust horse-power to indicated horse-power to be 60 per cent. (which appears to be the ascertained ratio for torpedo boats and ships of fine lines), the equivalent indicated horse-power for 31·01 knots is 1576.

The feed water supplied to the boiler was measured by a Siemens water meter previously calibrated under the working conditions, and found to be substantially correct. These measurements were made when running at a speed of 28 knots, and the consumption at 31·01 knots has been calculated from these measurements according to the known law between steam pressure and consumption, and by the observed steam pressures on the engines at the respective speeds. The consumption at 31·01 knots is approximately 25,000 lb. per hour, or 15·86 lb. per indicated horse-power. It should be observed that the assumption of the thrust horse-power being 60 per cent. of the indicated horse-power presupposes that the propellers are of the best form attainable, and should those now fitted be superseded by others of higher efficiency, as is possible, and, indeed, probable, then the figures of consumption per indicated horse-power will be correspondingly improved, and the speed of the boat increased.

The consumption of steam at 11·4 knots speed has been measured by meter, and found to be 2700 lb. per hour, or equivalent to a coal consumption of about 24·6 lb. per knot.

Conditions of running of Turbinia at 31·01 knots speed

Mean revolutions of engines	2100
Steam pressure in boiler	200 lb.
Steam pressure at engines	130 lb.
Vacuum at exhaust of engines	13½ lb.
Speed of boat	31·01 knots
Calculated thrust horse-power	946
Calculated indicated horse-power	1576
Consumption of steam, reduced to basis of 31·01 knots ...	25,000 lb.
Consumption of steam per indicated horse-power per hour	15·86 lb.
Total weight of machinery, including boiler, condensers, engines, auxiliaries, shafting, propellers, tanks, water in boiler, and hotwell, in working order	22 tons
Indicated horse-power per ton of total machinery... ...	72·1

Owing to adverse weather, these trials have been much delayed, and had finally to be made under unfavourable circumstances. They are, however, believed to be substantially accurate.

––––––––––

Mr Parsons, replying to the discussion, said that Mr Thornycroft's question as to how far the speed of the turbines could be reduced was one which could not be answered in a word, but in ships of larger beam it would be certainly possible to make some reduction in the speed; in fact, the number of revolutions could be lessened in direct proportion as the diameter of the turbine was increased. Down to 15 knots screws could be driven direct; below that gearing would have to be used, and then the turbine would be applicable to all classes of ships. He looked with distrust on gearing; the small gearing might be made to work, but the application of large geared wheels was a questionable device. The steam pressure could be as high as the boiler would give. It was quite possible to use superheated steam; the difficulty was in getting it. If the superheater were put in the uptake, the temperature of which varied widely, according to the state of the fire, etc., or from combustion taking place there, the amount of superheat would also vary greatly. The chief advantage in using the turbine was that by compounding the steam could be expanded down to a lower pressure than ordinarily used; he was able to work down to 1 lb. final pressure, and at the same time could work with very high initial pressure. That gave a very wide range of expansion, and consequent economy. If he were to work up to 300 lb. pressure, he would obtain still better results.

PRESIDENTIAL ADDRESS

The Institution of Junior Engineers*, November 3rd, 1899

The definition of the aims of the engineer as formulated in the charter of 1828 of our parent society the Institution of Civil Engineers as "the art of directing the great sources of Power in nature for the use and convenience of man", was, at that date, an accurate and concise description of those aims which are still ours to-day and may be similarly described.

A great change has, however, taken place in the qualifications of the engineer in the last seventy years, a great increase in the knowledge of the sources of Power in nature and also of the forces, conditions, and behaviour which underlie and constitute these sources of Power.

We now enjoy the legacy not only of a larger collection of natural facts, but what is of more importance, those facts have been tabulated and analysed by the systematic application of methods which have been initiated to deal with and to harmonise them, and by interaction the facts have developed theory, and theory has led up to and developed new unsuspected facts; and further, the knowledge and training acquired in the different branches of science and research have been of mutual assistance and have furthered the cause of general progress.

We find the engineer of the present time with a far wider range and depth of knowledge and with carefully compiled textbooks, rules and tables on almost every conceivable subject at his disposal, and to deal with all this his faculties must be more highly trained than formerly; true, he should have experience, but not the so-called rule-of-thumb experience which formerly sufficed to carry him through with credit in days of ruder method —a knowledge of simple facts. He should have familiarised himself with wide ranges of facts elaborated into concrete form and ranged in due proportion of importance as bearing upon questions with which he may have to deal. Engineering design is in many respects a work of art, and in these days the artist's materials have become more complex and the standard of perfection has advanced. The subject on which I have been chiefly engaged for some years has been that of motive power and more especially motive power obtained from steam in the steam engine; also closely connected with it are some applications of power which it may be interesting to describe.

We may first of all consider the great sources of power which Nature has placed within our reach, and also the extent to which we have been able to utilise them.

* Since 1902 The Junior Institution of Engineers.

In the present state of our solar system there are three chief sources from which motive power may be obtained, though only one of the three has so far been used for practical purposes.

Firstly, there is the energy of the tides produced by the energy of rotation of the Earth under the attraction of the Sun and Moon; this energy is drawn almost entirely from the Earth whose speed of rotation is being gradually diminished, and the energy is expended in fluid friction and heating of the ocean, but the great cost of utilising some of this power by impounding the water has so far precluded its useful application.

Secondly, there is the heat of the Earth whose temperature normally increases by about 1° F. for every 80 feet of depth below the surface, but in volcanic regions the heat is much more readily available. The great difficulty, however, to be met with is the low conductibility and specific heat of rock or molten lava.

Thirdly, there is the group of sources primarily derived from the heat of the Sun which constitute at the present time practically all the available sources of power.

The vast mass of the Sun, at an estimated temperature above 6000° F., radiates light and heat in all directions into space. This great flow consists of transverse vibrations in the ether and takes place in straight lines; of the whole, only about one two-thousand-millionth falls on the Earth's surface. Of this quantity of heat received nearly the whole is radiated away again from the Earth's surface into space, but from it in its course we derive our motive power.

The character of this flow of heat from a very hot source like the Sun to a much cooler recipient like the Earth, and its final rejection by the latter into space, is analogous to a waterfall from a high lake flowing to one of lower level, and is also analogous to the conditions under which all heat engines are placed, for were it possible to place a boiler at the Sun and lead a steam pipe to the Earth and condense in the Ocean we should have a steam engine without fuel, in fact could we only cause some of the vast flow of heat to pass through a heat engine on the Earth, we should secure a vast source of power; this has not yet been done on a practical scale.

However, indirectly it produces some power in other ways. For instance, the atmosphere of the Earth is unequally heated by the Sun's rays causing convection currents, the chief sources of winds.

The Ocean is evaporated near the equator and the vapour condenses nearer the poles causing rainfall, rivers, and water power.

Again, some of the heat stream is caught by vegetation in which it expends its energy by causing dissociation of the chemical constituents of the leaves and the carbonic acid gas drawn from the atmosphere, producing carbon and hydrocarbons and setting free the oxygen, so forming wood, coal or fuel. When the fuel is burnt it unites with the oxygen of the air giving out some of the heat that had been originally absorbed from the Sun.

Perhaps, with the exception of petroleum whose origin is not fully determined as to whether it is the product of the Sun's heat on vegetation, or of the heat of the Earth itself acting on carbonates in the presence of water, the heat of all fuel may be said to have been derived from the Sun's radiant heat.

It is a very remarkable fact, however, that an infinitesimal portion of the Sun's heat falling on the surface of the Earth is converted into an available form from which may be derived mechanical energy, whether in the winds, waterfalls, or the production of fuel.

An engine may eventually be discovered that will be actuated directly by the rays of the Sun, and should its principle of action be based on the second law of thermodynamics, its *modus operandi* will be the absorption of the heat from the Sun in the day, the production of mechanical energy, and the rejection of the residuum of heat to some equaliser or dissipater of heat such as the Ocean.

Engines of this description have often been proposed, and to enable us to realise that they are not very far removed from the region of practical possibilities, let us consider the experiment of a parabolic reflector of 24 inches diameter placed in the Sun on a bright summer's day. A thin piece of sheet iron placed in the focus is almost instantly melted, and if a toy steam engine be placed with the focus of the mirror on the boiler, it will raise steam in a few minutes, and enough steam will be generated to work it continuously.

Even in this crude form the problem does not seem hopeless but is rather a question of first cost and cost of maintenance.

Sir Robert Ball says: "The total heat received by the Earth from the Sun in any given time is that intercepted by its diametrical cross section, i.e. by the area of one of its great circles kept always perpendicular to the Sun's rays. The quantity of ice that would be melted annually on the circular plane by the solar rays would be a sheet having a thickness of 546 feet".

He then calculates the thickness of the belt of ice at the latitude of Egypt, i.e. 30° lat., that would be melted in the year, and arrives at the figure of 151 feet. He further says that this is on the assumption of no absorption of heat by the atmosphere, but though this is considerable, yet most of the heat reaches the Earth.

Calculating from these figures the thermal units, we find that on 1 square foot of Earth's surface in Egypt 150 cubic feet of ice would be melted annually, which is equivalent to 1,278,000 thermal units, or equal to the production of 1100 lb. of steam at 250 lb. pressure; on the assumption of 16 lb. of steam per horse-power hour in a good condensing engine it is equal to about 68 horse-power hours. All this work is done in daylight on an average of 12 hours per day, therefore 1 square foot will provide 5·66 horse-power days, or 64·4 square feet will provide on the average 1 horse-power for 12 working hours during the year. If we seek to concentrate the Sun's heat on a boiler by a concave mirror, we find that after allowing 50 per cent. for the losses in absorption in the atmosphere and reflection, a mirror 13 feet in diameter mounted like a heliostat and reflecting the Sun's heat on to a small boiler and working a condensing steam engine of the best make, the condenser being cooled by sea or lake water at 80° F., will produce 1 horse-power.

But, again, the good steam engine which we have assumed as a practical example converts only about one-eighth of the total heat given to the boiler into mechanical power.

If, however, an engine were discovered which would convert one-half the total heat received into mechanical work, a mirror of only 6 ft. 6 in. diameter would be required to produce 1 horse-power on the same basis, and it does not seem unreasonable to expect that such a motor may eventually be constructed.

All heat engines at present in use work under the second law of thermodynamics, that is to say they take in heat from a source at a high temperature and discharge most of it at a lower temperature, the disappearance of heat in the process being the equivalent of the work done by the engine. In all cases at the present time the source of heat is fuel of some kind, and after working the engine the residue is discharged in the case of the steam engine either to the condenser, or in the exhaust steam when non-condensing. In the gas engine it is discharged in the waste gases and into the water jacket around the cylinder.

The chief types of heat engine are:

(1) *Steam engines* of the turbine or reciprocating type which under favourable conditions generate one effective horse-power hour with $1\frac{3}{4}$ lb. of good coal, and whose complete weight with boiler in working order has been reduced to 20 lb. per horse-power, and in some cases for flying machines to 8 lb. per horse-power.

(2) *Gas engines* which in the larger sizes consume $1\frac{1}{3}$ lb. of anthracite coal or 17 cubic feet of illuminating gas per effective horse-power hour.

(3) *Oil engines* which consume in some cases 0·82 lb. of oil per effective horse-power hour, and whose weight for motor car propulsion has been reduced to $39\frac{1}{2}$ lb. per horse-power.

(4) *Hot-air engines* consuming from 2 to 8 lb. of coal per effective horse-power hour.

(5) *Gunpowder engines* or cannon, consuming in moderate sizes 4·1 lb. of cordite (having the same calorific value as 0·7 lb. of coal) per horse-power hour in the shot. The weight of a 6-inch quick-firing gun (without mounting) is 10 lb. per average effective horse-power in the shot at 6 rounds per minute. In the Maxim gun the weight is about 1 lb. per effective horse-power.

Primary batteries constitute a form of engine or producer of electrical energy that consumes fuel of a special kind prepared in a form suitable for assimilation; they do not work under the second law of thermodynamics, and the efficiency is high, but the great cost of their special fuel has restricted their use to very small powers.

The earliest records of heat engines are found in the "Pneumatics" of Hero of Alexandria about 200 B.C. He describes a reaction steam turbine, a spherical vessel mounted on an axis and supplied with steam through one of the trunnions from a boiler beneath. The steam escaping through two nozzles diametrically opposite to each other and tangential to the sphere, caused the sphere to rotate by the reaction or momentum of the issuing steam, in a manner analogous to a Barker's water wheel.

Thus, the first engine deriving its motive power from fuel was a crude form of steam turbine, and though it could have been applied to useful work and could easily have been made sufficiently economical to replace manual and horse-power in many instances, yet it lay dormant till A.D. 1629, when Branca suggested the same principle in a different form. Branca's steam turbine consisted simply of a steam jet fed from a boiler impinging against vanes or paddles attached to the rim of a wheel which was blown round by the momentum of the steam issuing from the jet.

The piston engine is, however, of comparatively modern origin and dates from about the year A.D. 1700. Engines of this class are so well known that it suffices to say that they have been practically the sole motive power engines from fuel in use from 1700 up to 1845, and have constituted one of the most important factors in the development of modern engineering enterprise.

Air engines were introduced about the year 1845 and although the larger engines of the Stirling type were very economical in fuel, yet on account of the inherent difficulty of heating large volumes of air within metal chambers or pipes, a difficulty arising from the low conductibility of air and consequently the overheating and burning of the metal, they have only come into commercial use for very small powers.

During the last thirty-five years gas engines have been perfected, and more recently oil engines, and in point of efficiency both convert a somewhat larger percentage of the heat energy of the fuel into mechanical energy than the best steam engines.

All successful oil and gas engines are at present internal combustion engines, the fuel being burnt in a gaseous form inside the working cylinder.

Very numerous attempts have, however, been made to construct internal combustion engines to burn solid fuel instead of gas. Some have been so far successful as to work with good economy in fuel, but the bar to their commercial success has been the cutting of the cylinder and valves by fine particles of fuel. This difficulty is not present when the fuel is introduced in the gaseous or liquid form, and hence the success of gas and oil engines, but could this difficulty be overcome the solid fuel would be the cheaper to use.

Internal combustion engines, gas engines, oil engines, cannon, etc., owe their superior economy in fuel to the very high temperature at which the heat is transferred from the fuel to the working substance of the engine, and consequently the great range of temperature in the working substance of the engine. In steam engines the temperature is limited by the practical difficulties of deterioration of metal and materials involved in the construction.

However, though this is the state of the case at the high end of the temperature range, in the engines we are considering the case is different at the lower end of the temperature range.

In all the internal combustion engines mentioned the consumed gases are rejected at a comparatively high temperature, while in the condensing steam engine the heat is rejected to the condenser at a comparatively low temperature. Internal combustion engines may be said to work between limits of temperature high up on the temperature scale and further removed from the absolute zero, while the steam engine works between temperatures lower down on the temperature scale and nearer the absolute zero.

As it is the ratio of the heights of the higher to the lower limits of temperature from the absolute zero of temperature that broadly speaking measures the economy of the engine, it follows that the internal combustion engine has the advantage at the high end, while the steam engine has the advantage at the low end, and, as we shall endeavour to show, steam engines of the turbine class have advantages over the piston engines in permitting a larger extension of the temperature range at the low end.

About fifteen years ago I was lead by circumstances to investigate the subject of improving the steam turbine. In recent times several attempts had been made to apply steam turbine wheels of the Hero and Branca types to the driving of circular saws and fans. The velocity of rotation with either of these types must necessarily be very high in order to obtain a reasonable efficiency from the steam, a velocity much in excess of that suitable for the direct driving of almost all classes of machinery; gearing was considered objectionable, and it therefore appeared desirable to adopt some form of turbine in which the steam should be gradually expanded in small steps or drops in pressure so as to keep the velocity of flow sufficiently low to allow of a comparatively moderate speed of rotation of the turbine engine.

The method adopted was to gather a number of turbines of the parallel flow type on to one shaft and contained in one case, the turbines each consisting of a ring of guide and a ring of moving blades, the successive rings of blades, or turbines, being graduated in size, those nearer the exhaust end being larger than those near the steam inlet, so as to allow a gradual expansion of the steam during its passage through the turbines.

The loss of power present in engines of the piston class due to cylinder condensation arising from the variation of steam pressure in the cylinder is not present in the steam turbine, as the steam pressure remains constant at each turbine ring and each part of the cylinder and barrel, and the numerous tests of steam consumption that have been made have shown that compound steam-turbine engines of moderate sizes when working with a condenser are comparable in steam consumption per effective horse-power with the best compound or triple, condensing, steam engines of the piston type. They have been constructed in sizes up to about 1000 horse-power for driving alternators and dynamos, and several sets of about 2000 horse-power are nearing completion. They are in use at the Newcastle, Cambridge, Scarborough, London Metropolitan, and other stations. Recently steam turbines have been successfully applied to driving centrifugal and screw fans and pumps, and as an instance of endurance, one turbo-screw fan for induced draught, of about 60 horse-power, has been at work continuously day and night for the last three years with the exception of Christmas and midsummer holidays.

There is no remarkable feature to note about centrifugal fans and pumps worked from steam turbines, excepting that in the case of pumps it has been found that with a slight modification of proportions lifts up to 200 feet can be dealt with in a single pump with the same efficiency of pump as in the case of ordinary centrifugal pumps at moderate lifts.

But the screw fans and pumps have some features of novelty which it may be interesting to mention. The runner or screw which drives the air or water is similar to a ship's propeller and is placed in a double truncated conical pipe, at the narrowest part between a trumpet-mouthed converging entry and a gently diverging conical exit. By this means the air or water is gently accelerated up to a suitable velocity at the propeller, and by the diverging cone the velocity is gradually checked, and the energy converted into available pressure to overcome the head against which the fan or pump has to work. Under most conditions the efficiency is found to be almost equal to that of the centrifugal fan with spiral case, and superior to the ordinary centrifugal open fans without casing, while the volume of air or water passed is much greater in proportion to the size. These results might naturally be expected, for on consideration it will be seen that the only losses are those due to friction in the conical pipes, and the ordinary losses of screw propellers.

The application of the compound steam turbine to the propulsion of vessels is a subject of considerable general interest in view of the possible and probable general adoption of this class of engine in fast vessels.

In the turbine is found an engine of extremely light weight, with a perfectly uniform turning moment, and very economical in steam in proportion to the power developed, and further it can be perfectly balanced so that no perceptible vibration is imparted to the ship. The problem of proportioning the engine to the screw propellers and to the ship to be driven has been the subject of costly experiments extending over several years; with the result that a satisfactory solution has been found giving very economical results in regard to pounds of steam consumed in the engines per effective horse-power developed in propelling the vessel, results which are equal or superior to those so far obtained with triple expansion engines of ordinary type in torpedo boats or torpedo-boat destroyers. The arrangements adopted may be best described by saying that instead of placing as usual one engine to drive one screw shaft, the turbine engine is divided into two, three or sometimes more, separate turbines, each driving a separate screw shaft, the steam passing successively through these turbines; thus when there are three turbines driving three shafts, the steam from the boiler passes through the high-pressure turbines, thence through the intermediate, and lastly through the low, and thence to the condenser.

As to the propellers, these approach closely to the usual form. It has, however, been found best to place two propellers of approximately the same pitch on each shaft at some considerable distance apart so that the after one shall not be seriously affected by the wash of the one in front. The advantage of this arrangement is that a sufficient blade area is obtained to carry the thrust necessary to drive the vessel with a lesser diameter of propeller, and so permitting of a higher speed of revolution of the engines.

The problem was complicated by the question of cavitation, which, though previously anticipated, was first practically found to exist by Mr Thornycroft and Mr Barnaby in 1894, and by them it was experimentally determined that cavitation or the hollowing out of the water into vacuous spaces and vortices by the blades of the propeller commences to take place when the mean thrust pressure on the projected area of the blades exceeds $11\frac{1}{4}$ lb. per square inch.

This limit has since been corroborated during the trials of the *Turbinia*.

This phenomenon has also been further investigated in the case of model propellers working in an oval tank of water, and to permit of cavitation at more moderate speeds than would otherwise have been necessary, the following arrangement was adopted: the tank was closed, plate-glass windows being provided on each side through which the propeller could be observed, and the atmospheric pressure was removed from the surface of the water by an air pump; under this condition the only forces tending to prevent cavitation were the small head of water above the propeller, and capillary attraction.

In the case of a propeller of 2 inches in diameter, cavitation commenced at about 1200 revolutions and became very pronounced at 1500. Had the atmospheric pressure not been removed, speeds of 12,000 and 15,000 respectively would have been necessary.

Photographs were taken with a camera made for the purpose with a focal plane shutter giving an exposure of about one-thousandth of a second, the illumination being by sunlight concentrated on the propeller by a 24-inch concave mirror.

Photographs were also taken by intermittent illumination of the propeller from an arc lamp, the arrangement consisting of an ordinary lantern condenser, which projected the beam on to a small concave mirror, mounted on a prolongation of the propeller shaft, the reflected beam being caught by a small stationary concave mirror at a definite position in each revolution and reflected on to the propeller. By this means the propeller was illuminated in a definite position at each revolution, and to the eye it appeared as stationary. The cavities about the blades could also be clearly seen and traced. Photographs were taken with an ordinary camera and about 10 seconds exposure.

A series of experiments was also made with model propellers in water at and just below the boiling point, dynamometric measurements being taken of power and thrust with various widths of propeller blade, the conclusion arrived at being that wide and thin blades are essential for fast speeds at sea, as well as a coarse pitch ratio of propeller.

The first vessel fitted with steam-turbine machinery was the *Turbinia*. She was commenced in 1894 and after many alterations and preliminary trials was satisfactorily completed in the spring of 1897.

The first vessels of larger size than the *Turbinia* to be fitted with steam-turbine machinery are the torpedo-boat destroyer *Viper* for our own Government, and a similar vessel for Messrs Sir W. G. Armstrong, Whitworth & Co.

These vessels are of approximately the same dimensions as the 30-knot destroyers now in Her Majesty's service, but have slightly more displacement. The boilers are about 12 per cent. larger, and it is estimated that upwards of 10,000 horse-power will be realised under the usual conditions as against 6500 with reciprocating engines.

The engines of these vessels are in duplicate. Two screw shafts are placed on each side of the vessel, driven respectively by a high- and a low-pressure turbine; to each of the low-pressure turbine shafts a small reversing turbine is permanently coupled for going astern, the estimated speed astern being $15\frac{1}{2}$ knots and ahead 35 knots; two propellers are placed on each shaft.

The latter of these two vessels has commenced her preliminary trials and has already reached a speed of upwards of 32 knots. The manipulation of the engines is a comparatively simple matter, as to reverse it is only necessary to close one valve and open another, and owing to there being no dead centres, small graduations of speed can be easily made.

In regard to the general application of turbine machinery to large ships, the conditions appear to be more favourable in the case of the faster class of vessels such as cross-channel boats, fast passenger ships, cruisers and liners; in such vessels the reduction in weight of machinery as well as economy in the consumption of coal per horse-power are important factors, and in some vessels the absence of vibration, both as regards the comfort of passengers, and in the case of ships of war, permitting of greater accuracy in sighting of the guns, is a question of first importance.

As regards cross-channel boats, the turbine system presents advantages in speed, absence of vibration, and, owing to the smaller diameter of the propellers, reduced draught of water.

As an instance, a boat of 270 feet length, 33 feet beam, 1000 tons displacement, and 8 ft. 6 in. draught of water could be constructed with spacious accommodation for 600 passengers, and with machinery developing 18,000 horse-power; she would have a sea speed of about 30 knots as compared with the speed of 19 to 22 knots of the present vessels of similar size and accommodation.

It is, perhaps, interesting to examine the possibilities of speed that might be attained in a special unarmoured cruiser, a magnified torpedo-boat destroyer of light build with scanty accommodation for her large crew, but equipped with an armament of light guns and torpedoes. Let us assume that her dimensions are about double those of the 30-knot destroyers, with plates of double the thickness and specially strengthened to correspond with the increased size; length 420 feet, beam 42 feet, maximum draught 14 feet, displacement 2800 tons, indicated horse-power 80,000. There would be two tiers of water tube boilers; these with the engine space, coal bunkers, etc., would occupy the whole of the lower portion of the vessel, the crews' quarters and guns would be on the upper decks. There would be eight propellers of 9 feet diameter revolving at about 400 revolutions per minute, and her speed would be about 44 knots.

She could carry coal at this speed for about 8 hours, but she would be able to steam at from 10 to 14 knots with a small section of the boilers more economically than other vessels of ordinary type and power, and

PLATE II

From a snapshot by Lady Parsons in South Africa, 1929

On board the *Caronia*, British Association Meeting in Canada, 1924 with Professor
J. C MacLennan, Lord Rutherford and Sir Thomas Holland

when required, all the boilers could be used, and full power exerted in about half an hour.

In the case of an Atlantic liner or a cruiser of large size, turbine engines would appear to present some considerable advantages. In the first place they would effect a reduction in weight of machinery, and some increase in economy of fuel per horse-power developed, both thus tending either to a saving in coal on the one hand, or, if preferred, some increase in speed.

The advantages are, however, less pronounced in this class of vessel on account of the smaller relative power of the machinery, and the large quantity of coal necessary for long voyages, but the complete absence of vibration at all speeds, not to mention many minor considerations of saving in cost and reduced engine-room staff are questions of considerable importance.

In conclusion, it may be remarked that in the history of engineering progress, the laws of natural selection generally operate in favour of those methods which are characterised by the greater simplicity and greater efficiency, whether these advantages be great or small.

MOTIVE POWER—HIGH-SPEED NAVIGATION— STEAM TURBINES

Royal Institution, Friday, January 26th, 1900

Twenty centuries ago the political power of Greece was broken, although Grecian civilisation had risen to its zenith. Rome was growing continually stronger, and was rapidly gaining territory by absorbing weaker states. Egypt, older in civilisation than either Greece or Rome, fell, but two centuries later, before the assault of the younger states, and became a Roman province. Her principal city at this time was Alexandria, a great and prosperous city, the centre of the commerce of the world, the home of students and of learned men, its population the wealthiest and most civilised of the then known world.

It is among the relics of that ancient Egyptian civilisation that we find the first records of the early history of the steam engine. In Alexandria, the home of Euclid, and possibly contemporary with Archimedes, Hero wrote his *Spiritalia seu Pneumatica*. It is doubtful if Hero was the inventor of the contrivances and apparatus described in his work; it is more probable that they were devices generally known at the time. Nothing in the text, however, indicates to whom the several machines are to be ascribed. Two of these machines are of special interest. The first utilised the expansive force of air in a closed vessel heated externally, the pneumatic force being applied upon the surface of water in other vessels, and the hydraulic force utilised for opening the doors of a Grecian temple and working other pseudo-magic contrivances.

Then after describing several forms of cylindrical boilers, and the use of the steam jet for accelerating combustion, he comes to the first of a type of steam engine, the steam turbine, which is the subject of our discourse this evening.

This is a veritable steam engine. The cauldron contains water, and is covered by a steam-tight cover, a globe is supported above the cauldron by a pair of tubes, one carrying a pivot, and the other opening directly through the trunnion joint into the sphere; short bent pipes are attached to diametrically opposite points on the equator. The steam generated in the cauldron passes up into the sphere and issues tangentially from the bent pipes, and by the reaction causes the sphere to rotate.

It seems uncertain whether this machine was ever more than a toy, or whether it was used by the Greek priests for producing motion of apparatus in their temple; but from our experience within the last twenty years it appears that, with some improvements in design and construction, it could have been applied to perform useful work at the date of Hero*, and further that, when so improved, it might have claimed a place among economical steam engines, even up to the middle of the present century.

A few years ago I had an engine constructed to test the capabilities of this class of reaction steam turbine, the only difference between this engine and Hero's being that the sphere was abolished, as a useless encumbrance, the arms were made of thin steel tube of oval form, so as to offer the least resistance to their motion, and the whole was enclosed in a cast-iron case which was connected to a condenser. When supplied with steam at a pressure of 100 lb. per square inch, and a vacuum in the case of 27 inches of mercury, a speed of 5000 revolutions per minute was attained, and an effective power was realised of 20 horse, and the consumption of steam was only 40 lb. per brake horse-power. By this very creditable performance, I was encouraged to further test the system, and constructed a compound reaction engine, in which the steam was caused to pass successively through three pairs of arms on one hollow shaft, each pair being contained in a separate compartment through which the shaft passed, suitable metallic packing preventing the passage of steam from one compartment to the next. The performance of this engine was, however, not superior to that of the single two-arm Hero's engine, for the simple reason that the excessive resistance to motion of the arms in the denser steam of the compartments more than neutralised the gain from the compound form. The performance of this engine was, however, sufficiently good to have it placed on a par with many ordinary steam engines in the middle of the present century.

The great barrier to the introduction of Hero's engine was undoubtedly the excessive speed of revolution necessary to obtain economical results, and with the crude state of mechanical engineering at that time, it would have been a matter of some difficulty to construct the turbine engine with sufficient accuracy of workmanship for satisfactory results, to say nothing of the necessary gearing for applying the power to ordinary useful purposes.

The next steam engine mentioned in history, which is capable of practical and useful development, is Branca's†. It is of the simplest form, a jet of steam from a steam boiler impinges on a paddle wheel and blows it round. This form of engine has since 1889 been developed by Dr De Laval, of Stockholm, with great ingenuity, and is extensively used for moderate powers on the Continent. The speed is, however, necessarily very high in order to obtain economy in steam, and spiral reduction gearing is used in order that the speed of revolution may be reduced for the application of the power. The improvements that have been made in Branca's steam turbine

* Circa 150 B.C.

† [An Italian Engineer and Architect; he died in 1629. Ed.].

by De Laval are firstly, the ordinary steam jet is replaced by a diverging conical jet, which permits of the expansion of the steam before it emerges from the jet, so transforming the potential energy of the high-pressure steam into kinetic energy of velocity in the direction of flow.

Secondly, the crude paddle wheel of Branca is replaced by a wheel of the strongest steel, fringed round the periphery with little cupped blades of steel, somewhat analogous to the buckets of a Pelton water wheel.

Lastly, the steel wheel is mounted on a long and somewhat elastic shaft, to allow of its easy and free motion, and on one extremity of this shaft is mounted the pinion of the spiral reduction gear.

The speeds of revolution of the steam wheels of De Laval's turbine are from 10,000 to 30,000 revolutions per minute, according to the size, involving peripheral speeds up to 1200 feet per second, or about one-half the speed of the projectile from a modern cannon. Such speeds are necessary to obtain power economically from the high-pressure steam jet, issuing from 3000 to 5000 feet per second as calculated by Rankine.

It is somewhat remarkable that not till a century after Branca, the piston or ordinary reciprocating engine made its first appearance, in about the year 1705*, and has since become one of the chief factors in the great mechanical and engineering growths of the last century. During this period the steam turbine seems to have been, practically speaking, neglected, which is somewhat remarkable in view of the numerous attempts of inventors to construct a rotary engine, attempts which had no practical results.

In the year 1884, the advent of the dynamo-electric machine, and development of mechanical and electrical engineering, created an increased demand for a good high-speed engine. Engineers were becoming more accustomed to high speeds of revolution, for the speed of dynamos was at this time from 1000 to 2000 revolutions per minute, of centrifugal pumps from 300 to 1500, and wood-working machinery from 3000 to 5000; and Sir Charles Wheatstone had made a tiny mirror revolve at a speed of 50,000 revolutions per minute for apparatus for measuring the velocity of light. The problem then presented itself of constructing a steam turbine, or ideal rotary engine, capable of working with good economy of steam at a moderate speed of revolution, and suitable for driving dynamos without the intervention of reduction gearing. To facilitate the problem, the dynamo was also considered with the view of raising its speed of revolution to the level of the lowest permissible speed of the turbine engine. In other words, to secure a successful combination, the turbine had to be run as slowly as possible, and the dynamo speed had to be raised as much as possible, and up to the same speed as the turbine, to permit direct coupling.

In 1884 preliminary experiments were commenced at Gateshead-on-

* [Newcomen. 1663–1729. Ed.]

Tyne, with the view of ascertaining by actual trial the conditions of working equilibrium and steady motion of shafts and bearings at the very high speeds of rotation that appeared to be essential to the construction of an economical steam turbine of moderate size. Trial shafts were run in bearings of different descriptions up to speeds of 40,000 revolutions per minute; these shafts were $1\frac{1}{2}$ inches in diameter and 2 feet long, the bearings being about $\frac{3}{8}$ inch in diameter. No difficulty was experienced in attaining this immense speed, provided that the bearings were designed to have a certain small amount of "give" or elasticity; and after the trial of many devices to secure these conditions, it was found that elasticity, combined with frictional resistance to transverse motion of the bearing bush, gave the best results, and tended to damp out vibrations in the revolving spindle. This result was achieved by a simple arrangement; the bearing in which the shaft revolved was a plain gun-metal bush with a collar at one end and a nut at the other; on this bush were threaded thin washers, each being alternately larger and smaller than its neighbour, the small series fitting the bush and the larger series fitting the hole in the bearing block, these washers occupying the greater part of the length of the bush. Lastly, a wide washer fitted both the bush and block, forming a fulcrum on which the bush rested; while a spiral spring between the washers and the nut on the bush pressed all the washers tightly against their neighbours. It will be seen now that, should the rotating shaft be slightly out of truth (which it is impossible to avoid in practice), the effect is to cause a slight lateral displacement of the bearing bush, which is resisted by the mutual sliding friction of each washer against its neighbour. The shaft itself being slightly elastic, tends to centre itself upon the fulcrum washer before mentioned, under the gyrostatic forces brought into play by the rapid revolutions of the shaft and influenced by the frictional resistance of the washers, and so the shaft tends to assume a steady state of revolution about its principal axis, or the axis of the mass, without wobbling or vibration. This form of bearing was exclusively used for some years in turbine engines aggregating some thousands of horse-power, but it has since been replaced by a simpler form fulfilling the same functions. In this later form the gun-metal bush is surrounded by several concentric tubes fitting easily within each other with a very slight lateral play; in the interstices between the tubes the oil enters, and its great viscosity when spread into thin films has the result of producing great frictional resistance to a rapid lateral displacement of the bearing bush; the oil film has also a centring action, and tends under vibration to assume a uniformity of thickness around the axis, thus centring the shaft, and like a cushion damping out vibrations arising from errors of balance. This form of bearing has been found to be very durable and quite satisfactory under all conditions.

Having tested the bearings up to speeds above those contemplated in the steam turbine, the next problem was the turbine itself. The laws regulating the flow of steam being well known (which was not the case in Hero's time), various forms of steam turbine were considered, and it appeared desirable to adopt in principle some type that had been both successful in the water turbine, and also easily adapted to a multiple or compound formation, a construction in which the steam should pass successively through a series of turbines one after the other.

The three best known of water turbines are the outward flow, the inward flow, and the parallel flow, and of these the latter appeared to be the best adapted for the multiple or compound steam turbine, for reasons which will afterwards appear.

The object in view being to obtain a good coefficient of efficiency from the steam with a moderate speed of revolution and diameter of turbine wheel, it becomes essential that the steam shall be caused to pass through a large number of successive turbines, with a small difference of pressure urging it through each individual turbine of the set, so that the velocity of flow of the steam may have the proper relation to the peripheral velocity of the turbine blades to secure the highest degree of efficiency from the steam, conditions analogous to those necessary for high efficiency in water turbines. A large diameter of turbine wheel, it is true, would secure a moderate speed of revolution, but this may be dismissed at once for the simple reason that the frictional resistance of such a disc revolving at the immense peripheral velocity, in the exhaust steam, would make it a most inefficient engine.

In the year 1884, a compound steam-turbine engine of 10 horse-power and a modified high-speed dynamo were designed and built for a working speed of 18,000 revolutions per minute. This machine proved to be practically successful, and subsequently ran for some years doing useful work, and is now in the South Kensington Museum.

In 1889, in consequence of partnership difficulties and the temporary loss of patents, the radial flow type of turbines was reluctantly adopted. This type of turbine consists of a series of fixed discs with interlocking flanges at the periphery, forming, when placed together coaxially, a cylindrical case, with inwardly projecting annular discs. On the shaft are keyed a similar set of discs, the faces of the fixed and moving disc lie a short distance apart. From the faces of the fixed disc project rows of guide blades which nearly touch the moving disc, and from the moving disc project rows of moving blades which nearly touch the fixed disc.

In 1892, this type was the first to be adapted to work in conjunction with a condenser. The first condensing turbine of the radial flow type was of 200 horse-power, and at a speed of 4800 revolutions per minute, drove an alternator of 150 kilowatts output. It was tested by Professor Ewing *, and the general result of the trials was to demonstrate that the condensing steam turbine was an exceptionally economical heat engine. With a steam pressure of 100 lb., the steam being moderately superheated, and a vacuum of 28 inches of mercury, the consumption was 27 lb. per kilowatt hour, which is equivalent to about 16 lb. of steam per indicated horse-power. This result marked an era in the development of the steam turbine, and opened for it a wide field, including some of the chief applications of motive power from steam. At this period turbine alternators of the condensing type were placed in the Newcastle, Cambridge, and Scarborough Electric Supply Company's Stations, and soon afterwards several of 600 horse-power of the non-condensing parallel flow type were set to work in the Metropolitan Companies' Stations, where the comparative absence of vibration was an important factor. Turbine alternators and turbine dynamos of 2500 horse-power are now in course of construction in England and the United States, and larger sizes are in prospect.

A turbo-alternator manufactured at Heaton Works, Newcastle-on-Tyne, for the Corporation of Elberfeld in Germany, was tested a few days ago by a committee of experts from Germany, Professor Ewing being also present, with the following remarkable results. At the full load of 1200 kilowatts, and with a steam pressure of 130 lb. at the engine, and 10° C. of superheat, the engine driving its own air pumps, the consumption of steam was found to be at the rate of 18·8 lb. per kilowatt hour. To compare this figure with those obtained with ordinary piston engines of the highest recorded efficiencies, and assuming the highest record with which I am acquainted of the ratio of electrical output to the power indicated in the steam engine, namely 85 per cent., the figure of 18·8 lb. per kilowatt in the turbine plant is equivalent to a consumption of 11·9 lb. per indicated horse-power, a result surpassing the records of the best steam engines in the production of electricity from steam.

Turbine engines are also used for generating electrical current for the transmission of power, the working of electrical tramways, electrical pumping and similar purposes. They are also used for coupling directly to and driving fans for producing forced and induced draught for general ventilating purposes, also for driving centrifugal pumps for lifts up to 200 feet, and screw pumps for low lifts.

The most important field, however, for the steam turbine is undoubtedly in the propulsion of ships. The large and increasing amount of horse-power and the greater size and speed of the modern engines tend towards some form which shall be light, capable of perfect balancing and economical in steam. The marine engine of the piston type does not entirely fulfil all these

* Now Sir James Alfred Ewing, K.C.B.

requirements, but the compound turbine engine, as made in 1892, appeared to be capable of doing so, and of becoming an ideal marine engine. On the other hand, an element of uncertainty lay in the high speed of the turbine engine, and to couple it directly to a propeller of ordinary proportions would have led to failure.

In January, 1894, a pioneer syndicate was formed to explore the problem, those chiefly associated in the undertaking being the Earl of Rosse, Christopher Leyland, John Simpson, Campbell Swinton, Norman Cookson, the late George Clayton, H. C. Harvey, and Gerald Stoney. It was deemed expedient, for reasons of economy and also of time (as many alterations were anticipated), to build as small a vessel as possible, but not so small as to preclude the attainment of an unprecedented high speed in the event of success. The *Turbinia* was constructed, her dimensions being 100 feet in length, 9 feet beam, 3 feet draught of hull, and 44 tons displacement. She was fitted with a turbine engine of 2000 actual horse-power, with an expansion ratio of a 150-fold, also with a water-tube boiler of great power, of the express type, with small tubes. The turbine engine was designed to drive one screw shaft at a speed of from 2000 to 3000 revolutions per minute.

Many trials were made with screw propellers of various sizes and proportions, but the best speeds were quite disappointing, and it was clear that some radical defect lay in the propellers. This was corroborated by dynamometric measurements. The excessive slip of the propellers beyond the calculated amount, and their inefficiency, indicated a want of sufficient blade area upon which the thrust necessary to drive the ship was distributed —in other words, the water was torn into cavities behind the blades. These cavities contained no air, but only vapour of water, and the greater portion of the power of the engine was consumed in the formation and maintenance of these cavities instead of the propulsion of the vessel. This phenomenon was first noticed in the trials of the torpedo boat *Daring*, by Messrs Thornycroft and Mr Barnaby, shortly before the commencement of the trials of the *Turbinia*, and was named "cavitation" by Mr R. E. Froude.

This phenomenon has been investigated experimentally with propellers of small size working inside an oval tank, so as to represent approximately the conditions of slip ratio customary in fast ships. To enable the propeller to cause cavitation more easily the tank is closed and the atmospheric pressure removed from the surface of the water above the propeller by an air pump, glass windows are fitted for observation and illumination. Under these conditions the only forces tending to hold the water together and resist cavitation are the small head of water above the propeller, and capillarity. The propeller is 2 inches diameter and 3 inches pitch; cavitation commences at about 1200 revolutions and becomes very pronounced at

PLATE III

The *Turbinia* at 35 knots

H.M. Destroyer *Viper* at 37 knots

1500 revolutions. Had the atmospheric pressure not been removed, speeds of 12,000 and 15,000 revolutions per minute would have been necessary, rendering observations more difficult. The inference to be drawn from these experiments seems to be that for fast speeds of vessels, wide thin blades, a coarse pitch ratio, and moderate slip, are desirable for the prevention of cavitation, and in order to obtain the best efficiency in propulsion of the vessel.

To return to the *Turbinia*, a radical alteration was deemed necessary. A new turbine engine was made, consisting of three separate engines—high pressure, intermediate pressure, and low pressure—each of which drove one screw shaft, so that the power of the engine was distributed over three shafts instead of concentrated on one, and three propellers were placed on each shaft. The result of these changes was marvellous. The vessel now nearly doubled her speed, 30 knots was soon reached, and finally 32¾ knots mean speed on the measured mile authenticated, or the fastest speed then attained by any vessel afloat. The economy of her engines was investigated by Professor Ewing, assisted by Professor Dunkerly: the consumption of steam per indicated horse-power for all purposes at 31 knots speed was found to be 14½ lb., or in other words, with a good marine boiler the coal consumption would be considerably under 2 lb. per indicated horse-power, a result better than is obtained in torpedo-boats or torpedo-boat destroyers with ordinary triple expansion engines.

The vessel's reversing turbine gave her an astern speed of 6½ knots, and she could be brought to rest in 36 seconds when running at 30 knots speed, and from rest she could be brought up to 30 knots in 40 seconds.

The *Turbinia* cruised from the Tyne to the Naval Review at Spithead, where she steamed on the day of the Review at an estimated speed of 34½ knots. These results represent about 2300 indicated horse-power, and may be said to have been obtained without a very abnormal performance as regards the boiler; its total heating surface being 1100 square feet, with an evaporation of about 28 lb. per square foot at the speed of 34½ knots.

These speeds were not obtained by bottling up the steam and opening the regulating valve on coming to the measured mile, but were maintained for many miles together with constant steam pressure, and as long as the fires were clean. On the other hand, the endurance of the engines themselves seems to be unlimited, all heavy pressures, including the thrust of the propellers, that would in ordinary engines come on the bearings, being counterbalanced by the steam pressure acting on the turbines.

It seems clear that the results obtained in the case of the *Turbinia* were almost entirely due to the economy in steam of the turbine engines, and the unusually small weight of the engines, shafting and propellers, in proportion to the power developed.

It may also be said that generally speaking every part of the machinery was as substantial as in naval vessels of the torpedo-boat class, yet she developed 100 horse-power per ton of machinery, and 50 horse-power per ton of total weight of vessel in working order.

The results of the *Turbinia* having been found satisfactory, the original company which built her was merged into a large company under the same directorate for carrying on the work on a commercial scale. At Wallsend-on-Tyne, the Parsons Marine Steam Turbine Company erected works, and in 1898 contracted with the Admiralty for a 31-knot torpedo-boat destroyer, the *Viper* (Plate III), which is of the same dimensions as the usual 30-knot vessels of this class, viz. 210 feet length, 21 feet beam, and about 350 tons displacement, but with machinery of much greater power than usual in vessels of this size; they also contracted with Sir W. G. Armstrong, Whitworth and Co. for machinery for one of their torpedo-boat destroyers.

The turbine engines of these vessels are similar to those of the *Turbinia*, but are in duplicate, and consist of two distinct sets of engines on each side of the vessel. There are four screw shafts in all, entirely independent of each other, the two on each side being driven by one high and one low-pressure turbine respectively of about equal power; the two low-pressure turbines drive the two inner shafts, and to each a small reversing turbine is also permanently coupled, and revolves idly with them when going ahead. The screw shafts are carried by brackets as usual, and two propellers are placed on each shaft, the foremost in each case having a slightly lesser pitch than the after one. The thrust from the screw shafts is entirely balanced by the steam acting on the turbines, so that there is extremely little friction.

The boilers, auxiliary machinery and condensers are of the usual type in such vessels, but their size is somewhat increased to meet the much larger horse-power to be developed, and to compensate for the lesser weight of the main engines, shafting, propellers, as well as the lighter structure of the engine beds. The boilers are of the Yarrow type, with a total heating surface of 15,000 square feet, and grate surface of 272 square feet, and the condensers have a cooling surface of 8000 square feet. The hull and all fittings are of the usual design.

Let us consider the machinery (Plate IV) on one side of the vessel only: the steam from the boilers is admitted directly through a regulating valve to the high-pressure turbine driving one shaft, it then passes to the adjacent low-pressure turbine, driving its shaft independently, thence it flows to the condenser, and both the shafts then drive the vessel ahead; the reversing turbine revolves with the low-pressure shaft, and being permanently connected with the vacuum of the condenser no appreciable resistance is offered to its motion under these conditions. To go astern the ahead steam valve is closed and the astern valve opened, admitting the steam from the boilers to the reversing turbine, and reversing the direction of rotation of the inner screw shaft.

On the other side of the vessel the arrangement is the same, and it will be seen that she can be manœuvred as an ordinary twin-screw vessel, and with great facility and quickness.

On her second preliminary trial about three weeks ago, the mean speed of four consecutive runs on the measured mile reached 34·8 knots, and the fastest run was at the speed of 35·503 knots, which is believed to be considerably beyond the recorded speed of any vessel hitherto built. The vessel was scarcely completed at the time of this trial, and it is anticipated that still higher speeds will be realised on subsequent and official trials.[1] The speed of 35·5 knots, or nearly 41 statute miles, represents about 11,000 indicated horse-power in a vessel of 350 tons displacement, as compared with 6000 to 6500 developed in the 30-knot destroyers of similar dimensions and 310 tons displacement.

At all speeds there was very little vibration. Her speed astern is guaranteed to be 15½ knots.

The *Viper* has surpassed the *Turbinia* in speed, and is at the present time the fastest vessel afloat.

[1] The *Viper* has since attained with full trial weights on board a mean speed of 36·58 knots, on a 1-hour's full-power trial, the fastest runs being at the rate of 37·118 knots per hour.

THE STEAM TURBINE AND ITS APPLICATION TO THE PROPULSION OF VESSELS

Institution of Naval Architects, June 26th, 1903

[This paper opens with brief descriptions of steam turbines and passes on to accounts of the *Turbinia*, and H.M. Destroyers *Viper* and *Cobra*. The two destroyers named were lost in 1901. On August 3rd of that year the *Viper*, in a dense fog, ran ashore on Renouquet Island near Alderney and became a total wreck, and on September 18th the *Cobra* broke her back in very rough weather off the Outer Dowsing Shoal on the coast of Lincolnshire. A few words are said to dispel any lingering belief that these fatalities were in any way due to the turbine propelling machinery. The paper then continues as follows.]

More could be said as to the absence of reciprocating forces to shake the structure of the hull, but this seems unnecessary; and we pass on to consider the first passenger vessel to be propelled by steam turbines—the *King Edward*, built in the spring of 1901 by Messrs W. Denny and Brothers, of Dumbarton, and engined by the Parsons Marine Steam Turbine Company, Limited, to the order of Captain John Williamson, of Glasgow. She is 250 feet in length and 30 feet beam, with about 6 feet draught of water. The turbines are similar in construction to those of the *Turbinia*, and consist of three turbines—one high pressure, driving the centre shaft, and two low pressure, working in parallel, driving the side shafts; in the exhaust casing of each of the low-pressure turbines is placed the reversing turbine. On the centre shaft is one propeller 57 inches in diameter, and on each of the side shafts are two propellers of 40 inches diameter and about 9 feet apart. When manœuvring, steam is admitted by valves directly into the low-pressure, or, alternatively, to the reversing turbine for going ahead or astern on either side of the vessel, the centre shaft and turbine meantime remaining idle; when the main stop valve is opened to the high-pressure turbine, all the turbines go ahead. The auxiliary machinery is of the usual type, and needs no special mention beyond that the air pumps are worked by worm wheels from the low-pressure turbine shafts. The boiler is of the usual return-tube and double-ended type, for 150 lb. working pressure.

The trial of the *King Edward* was made on June 26th, 1901, on the Clyde, and on the Skelmorlie mile a mean speed of 20·48 knots was recorded, the revolutions of the centre shaft being 505 and of the side shafts 755.

The indicated horse-power was estimated to be 3500 from model experiments in the tank at Dumbarton. The average sea speed on the run of about 160 miles to Campbeltown and back in 1901 season was 19 knots, and the average coal consumption including lighting-up, etc., was 18 tons per day, or 1·8 lb. per equivalent indicated horse-power hour. The results of speed and coal consumption have been stated by Messrs Denny to be more favourable than those that could have been obtained from a similar vessel with triple-expansion reciprocating engines.

An interesting statement has been made by Mr James Denny to the effect that if the *King Edward* had been fitted with balanced twin triple-expansion engines of the most improved type, and of such size as would consume all the steam the existing boiler could make, the best speed that they could possibly expect would be 19·7 knots, as against the 20½ knots actually attained by the *King Edward*. The difference between 19·7 knots and 20½ knots corresponds to a gain in indicated horse-power in favour of the turbine steamer of 20 per cent. The performance of the vessel during the seasons of 1901 and 1902 has been most satisfactory, no hitch or trouble having been experienced from the machinery.

In 1902 the second turbine passenger vessel, the *Queen Alexandra*, was built by Messrs W. Denny and Brothers for Captain John Williamson, and engined by the Parsons Marine Steam Turbine Company, Limited. Her dimensions are 270 feet in length, 32 feet beam, and about 6 ft. 6 in. draught. Her machinery is similar to, but more powerful than, that of the *King Edward*. A change was also made in the working of the air pumps from the circulating engines instead of from the turbines.

The *Queen Alexandra* has a slightly larger boiler than the *King Edward*, but the steam pressure is the same—viz. 150 lb. The *Queen Alexandra* attained on trial a mean speed of 21·43 knots; the revolutions of the centre shaft were 750, and of the side shafts 1090. The indicated horse-power, taken from Messrs Denny's tank experiments, was estimated to be 4400, and the consumption of steam was under 15 lb. per indicated horse-power. In the spring of this year the tandem side propellers were replaced by single propellers of larger diameter and somewhat finer pitch, which has had a beneficial result in greater smoothness of motion, and a slight improvement in the speed and coal consumption. The success of these two Clyde vessels has led to the adoption of turbines for the latest addition to the Dover-Calais and the Newhaven-Dieppe services, and these fine vessels constitute a very important step in advance for marine steam-turbine machinery.

The Dover-Calais boat *Queen*, for the South Eastern and Chatham Railway Company, is 310 feet in length, 40 feet in beam, and 25 feet in depth. The machinery is designed for 8000 indicated horse-power on service. The trial of the *Queen* took place on the 12th of this month, on the Skelmorlie mile, when the mean speed of 21·73 knots was easily attained, which was considerably over the speed guaranteed by Messrs W. Denny and Brothers. When steaming continuously astern, the mean speed of 12·95 knots was easily obtained. Stopping and starting trials were also carried out, and the vessel, from a speed of slightly over 19 knots, was brought to a dead stop in 1 minute 7 seconds, and she travelled during this time 130 fathoms, or two-and-a-half times her own length. The Newhaven-Dieppe boat, for the London, Brighton and South Coast Railway, is 280 feet in length, 34 feet beam, and 22 feet in depth. This vessel was launched on the 13th of this month, and it is to be expected that she will be ready for her trials in a few weeks.

Three yachts have lately been fitted with steam turbines. The largest is the *Lorena*, built by Messrs Ramage and Ferguson, Limited, of Leith, to the order of Mr A. L. Barber, of New York, and to the designs of Messrs Cox and King, of London. This vessel is 253 feet in length on water-line, 33·3 feet beam, and about 13 feet draught of water. The turbines are similar to those of the *King Edward* and *Queen Alexandra*, but somewhat larger, and designed for a speed of vessel of 16 knots. She is fitted with four single-ended Scotch boilers for 180 lb. pressure, with a total heating surface of 8560 square feet, and a grate surface of 217 square feet, with Howden's forced draught. The adoption of turbines instead of reciprocating engines, as originally contemplated, permitted of a considerable increase of the saloon accommodation, besides saving 70 tons in the weight of the machinery.

The *Lorena's* trial took place in Aberlady Bay, in the Firth of Forth, on May 16th, and after several runs over the measured mile, the mean speed attained was 18·02 knots. The number of revolutions of the centre shaft was 550, and of the side shafts 700. The machinery was kept running at full speed for over five hours, and worked with the utmost smoothness. From observations taken on the trial, it was found that she will be very economical in coal consumption, and further trials of coal and water consumption are contemplated previous to her leaving for New York. The above full-speed trial was run with the vessel in normal cruising conditions at sea with about 240 tons of coal on board, or about half bunkers.

The steam yacht *Emerald*, built by Messrs Alex. Stephen and Sons, Limited, of Glasgow, to the order of Sir Christopher Furness, M.P., has been the first vessel propelled by turbines to cross the Atlantic. Her dimensions are: length, 198 feet; beam, 28 ft. 7 in.; and displacement about 900 tons. Her turbine machinery is similar to that of the two Clyde vessels, and she has one Scotch boiler with Howden's forced draught. She is fitted with one propeller on each shaft. She attained a mean speed of 15 knots on trial. The smoothness of running of the machinery and propellers is a noticeable feature. In crossing the Atlantic at the end of April she experienced very severe weather and head winds, but the propellers never raced, and the machinery worked admirably and without the slightest hitch.

The turbine yacht *Tarantula*, built by Messrs Yarrow for the late Colonel McCalmont, is similar in dimensions to a British first-class torpedo boat, but of heavier scantlings. Her turbine machinery is similar to that of the *Turbinia*, but she has in addition a cruising turbine to increase the economy at speeds up to 15 knots. On trial, she reached a mean speed of 25·36 knots on a displacement of 150 tons. This speed could probably have been increased by the substitution of single instead of tandem propellers, but she passed out of our hands before the probable benefit of such a change had been ascertained.

The *Velox*, a torpedo-boat destroyer built by Messrs Hawthorn, Leslie, and Co., to the order of the Parsons Marine Steam Turbine Company, Limited, and recently purchased by the British Admiralty, is fitted with machinery of the same power as that in the *Viper*, and, on emergency, capable of developing upwards of 10,000 indicated horse-power. A noticeable feature in this vessel is that two triple-expansion reciprocating engines are fitted, of 150 indicated horse-power each, connected by detachable claw couplings to the low-pressure turbine shafts. At cruising speeds up to 13 knots, steam, after working these engines, exhausts into the turbines, where its expansion is completed before passing to the condensers. The displacement when fully laden on trials was 440 tons. The consumption of coal at 27 knots worked out at 2·501 lb. per indicated horse-power, and at 11¼ knots 8½ cwt. per hour. At a speed of 31 knots, the estimated consumption would be about 2·3 lb. per indicated horse-power.

The turbine machinery on order for the *Eden* torpedo-boat destroyer, of 7000 indicated horse-power, and 25½ knots speed, and for the *Amethyst*, third-class cruiser, of 9800 indicated horse-power, and 21¾ knots speed, is similar, in general design and arrangement, to the engines of the passenger vessels previously described. But, in view of the great variation in horse-power required in modern war vessels, two additional cruising turbines are permanently coupled to the shafts of the main low-pressure turbines. When working at reduced power, the steam from the boilers passes through the cruising turbines in series, and thence to the main turbines; by this means a high ratio of expansion of the steam at all the lower speeds is obtained, and the loss by throttling of the steam avoided; suitable by-passes and admission valves being provided to the successive turbines on the down grade of the expansion curve. A by-pass admitting steam to the second stage of the expansion in the main high-pressure turbine provides means for developing an abnormal power when required, without a material loss of efficiency under such circumstances.

The very slow progress of the marine steam turbine at a time when the turbine was making rapid strides on land may seem somewhat remarkable, but it has been due to causes sufficiently obvious to all who have given attention to the subject. The Turbine Company, in 1897, directed their special and almost exclusive attention to the introduction of the turbine in destroyers as the most suitable and favourable application among all classes of vessels. Their efforts met with remarkable success in the direction of increased speed and diminished vibration, and immunity from breakdown, and good economy at the higher rates of speeds, but the very great delays and the immense difficulties and expense in carrying through such contracts, incidental to the introduction of a new class of machinery, and concluding by the total loss of the only two sets of machinery made at this period (other than the original *Turbinia*), had a very detrimental effect on the introduction of turbine machinery generally.

To-day, after six years of unremitting work by a highly skilled and devoted staff, and the expenditure of much money by the company, there is no turbine-propelled vessel actually in commission in H.M. service, though two turbine vessels are building, and one nearly completed vessel—the *Velox*—is being purchased for the Service. Both the *Viper* and *Cobra* were unfortunately lost before substantial practical experience in commission had been obtained. During and after the occurrence of these events, the Turbine Company directed their attention to the important application of their system to vessels of moderate speeds, and their efforts in this direction have met with much encouragement and great success. The names of Captain John Williamson, the managing owner, and Messrs Denny, the builders, will ever be associated with the *King Edward*, the first passenger vessel to be propelled by steam turbines. The vessels *King Edward, Queen Alexandra, The Queen,* and the yachts *Lorena, Emerald, Tarantula,* are now in commission and have given entire satisfaction to the owners and the public.

The engining of larger vessels and liners is not a very long step beyond what has already been proved to be successful. The experience with the marine turbine up to 10,000 horse-power in ships of fast as well as moderate speed has tended to justify the anticipation guided by theory, that the larger the engines the more favourable will be the results as compared with reciprocating engines. The saving of weight, cost, space, attendance, and upkeep will become still more marked with turbine engines of above 10,000 and up to 60,000 horse-power, for which designs have been prepared.

PRESIDENTIAL ADDRESS TO THE ENGINEERING SECTION

British Association, Cambridge, 1904

On this occasion I propose to devote my remarks to the subject of invention.

It is a subject of considerable importance, not only to engineers, but also to men of science and the public generally.

I also propose to treat invention in its wider sense, and to include under the word discoveries in physics, mechanics, chemistry, and geology.

Invention throughout the Middle Ages was held in little esteem. In most dictionaries it receives scant reference except as applied to poetry, painting, and sculpture.

Shakespeare and Dryden describe invention as a kind of muse or inspiration in relation to the arts, and when taken in its general sense to be associated with deceit as "Return with an invention, and clap upon you two or three plausible lies".

As to the opposition and hostility to scientific research, discovery, and mechanical invention in the past, and until comparatively recent times, there can be no question, in some cases the opposition actually amounting to persecution and cruelty.

The change in public opinion has been gradual. The great inventions of the last century in science and the arts have resulted in a large increase of knowledge and the powers of man to harness the forces of Nature. These great inventions have proved without question that the inventors in the past have, in the widest sense, been among the greatest benefactors of the human race. Yet the lot of the inventor until recent years has been exceptionally trying, and even in our time I scarcely think that anyone would venture to describe it as altogether a happy one. The hostility and opposition which the inventor suffered in the Middle Ages have certainly been removed, but he still labours under serious disability in many respects under law as compared with other sections of the community. The change of public feeling in favour of discovery and invention has progressed with rapidity during the last century. Not only have private individuals devoted more time and money to the work, but societies, institutions, colleges, municipalities, and Governments have founded many research laboratories, and in some instances have provided large endowments. These measures have increased the number of persons trained to scientific methods, and also provided greatly improved facilities for research; but perhaps one of the most important results to engineers has been the direct and indirect influence of the more general application of scientific methods to engineering.

Sir Frederick Bramwell, in his Presidential Address to this Association in 1888, emphasised the interdependence of the scientist and the civil engineer, and described how the work of the latter has been largely based on the discoveries of the former; while the work of the engineer often provides data and adds a stimulus to the researches of the scientist. And I think his remarks might be further appropriately extended by adding that since the scientist, the engineer, the chemist, the metallurgist, the geologist, all seek to unravel and to compass the secrets of Nature, they are all to a great extent interdependent on each other.

But though research laboratories are the chief centre of scientific invention, and colleges, institutions, and schools train the mind to scientific methods of attack, yet in mechanical, civil, and electrical engineering the chief work of practical investigation has been carried on by individual engineers, or by firms, syndicates, and companies. These not only have adapted discoveries made by scientists to commercial uses, but also in many instances have themselves made such discoveries or inventions.

To return to the subject, let us for a moment consider in what invention really consists, and let us dismiss from our minds the very common conception which is given in dictionaries and encyclopaedias that invention is a happy thought occurring to an inventive mind. Such a conception would give us an entirely erroneous idea of the formation of the great steps in advance in science and engineering that have been made during the last century; and, further, it would lead us to forget the fact that almost all important inventions have been the result of long training and laborious research and long-continued labour. Generally, what is usually called an invention is the work of many individuals, each one adding something to the work of his predecessors, each one suggesting something to overcome some difficulty, trying many things, testing them when possible, rejecting the failures, retaining the best, and by a process of gradual selection arriving at the most perfect method of accomplishing the end in view.

This is the usual process by which inventions are made.

Then after the invention, which we will suppose is the successful attempt to unravel some secret of Nature, or some mechanical or other problem, there follows in many cases the perfecting of the invention for general use, the realisation of the advance or its introduction commercially; this after-work often involves as great difficulties and requires for its accomplishment as great a measure of skill as the invention itself, of which it may be considered in many cases as forming a part.

If the invention, as is often the case, competes with or is intended to supersede some older method, then there is a struggle for existence between the two. This state of things has been well described by Mr Fletcher Moulton. The new invention, like a young sapling in a dense forest, struggles to grow up to maturity, but the dense shade of the older and higher trees robs it of the necessary light. If it could only once grow as tall as the rest all would be easy, it would then get its fair share of light and sunshine. Thus it often occurs in the history of inventions that the surroundings are not favourable when the first attack is made, and that subsequently it is repeated by different persons, and finally under different circumstances it may eventually succeed and become established.

We may take in illustration almost any of the great inventions of un-doubted utility of which we happen to have the full history—for instance, some of the great scientific discoveries, or some of the great mechanical inventions, such as the steam engine, the gas engine, the steamship, the locomotive, the motor car, or some of the great chemical or metallurgical discoveries Are not most, if not all, of these the result of the long-continued labour of many persons, and has not the financial side been, in most cases, a very important factor in securing success?

The history of the steam-engine might be selected, but I prefer on this occasion to take the internal-combustion engine, for two reasons—firstly, because its history is a typical one; and secondly, because we are to hear a paper by that able exponent and great inventor in the domain of the gas engine, Mr Dugald Clerk, describing not only the history, but the engine in its present state of development and perfection, an engine which is able to convert the greatest percentage of heat units in the fuel into mechanical work, excepting only, as far as we at present know, the voltaic battery and living organisms.

The first true internal-combustion engine was undoubtedly the cannon, and the use in it of combustible powder for giving energy to the shot is strictly analogous to the use of the explosive mixture of gas or oil and air as at present in use in all internal-combustion engines; thus the first internal-combustion engine depended on the combination of a chemical discovery and a mechanical invention, the invention of gunpowder and the invention of the cannon.

In 1680 Huygens proposed to use gunpowder for obtaining motive power in an engine. Papin, in 1690, continued Huygens' experiments, but without success. These two inventors, instead of following the method of burning the powder under pressure, as in the cannon, adopted, in ignorance of thermodynamic laws, an erroneous course. They exploded a small quantity of gunpowder in a large vessel with escape valves, which after the explosion caused a partial vacuum to remain in the vessel. This partial vacuum was then used to actuate a piston or engine and perform useful work. Subse-quently several other inventors worked on the same lines, but all of these failed on account of two causes which now are very evident to us. Firstly, gunpowder was then, as it still is, a very expensive form of fuel, in proportion to the energy liberated on explosion; secondly, the method of burning the powder to cause a vacuum involves the waste of nearly the whole of the available energy, whereas had it been burned under pressure, as in the cannon, a comparatively large percentage of the energy would have been converted into useful work. But even with this alteration, and however perfect the engine had been, the cost of explosives would have debarred its coming into use, except for very special purposes.

We come a century later to the first real gas engine. Street, in 1794, proposed the use of vapour of turpentine in an engine on methods closely analogous to those successfully adopted in the Lenoir gas engine of eighty years later, or thirty years ago. But Street's engine failed from crude and faulty construction. Brown, in 1823, tried Huygens' vacuum method, using fuel to expand air instead of gunpowder, but he also failed, probably on account of the wastefulness of the method.

Wright, in 1833, made a really good gas engine, having many of the essential features of some of the gas engines of the present day, such as separate gas and water pumps, and water-jacketed cylinder and piston.

Barnett, in 1839, further improved on Wright's design, and made the greatest advance of any worker in gas engines. He added the fundamental improvements of compression of the explosive mixture before combustion, and he devised means of lighting the mixture under pressure, and his engine conformed closely to the present-day practice as regards fundamental details. No doubt Barnett's engine, so perfect in principle, deserved commercial success, but either his mechanical skill or his financial resources were inadequate to the task, and the character of the patents would seem to favour this conclusion, both as regards Barnett and other workers at this period. Up to 1850 the workers were few, but as time went on they gradually increased in numbers; attention had been attracted to the subject, and men with greater powers and resources appear to have taken the problem in hand. Among these numerous workers came Lenoir, in 1860, who, adopting the inferior type of non-compression engine, made it a commercial success by his superior mechanical skill and resources. Mr Dugald Clerk tells us: "The proposals of Brown (1823), Wright (1833), Barnett (1838), Bansanti and Matteucci (1857), show gradually increasing knowledge of detail and the difficulties to be overcome, all leading to the first practicable engine in 1866, the Lenoir". This stage of the development being reached, the names of Siemens, Beau de Rochas, Otto, Simon, Dugald Clerk, Priestman, Daimler, Dowson, Mond, and others, appear as inventors who have worked at and added something to perfect the internal-combustion engine and its fuel, and who have helped to bring it to its present state of perfection.

In the history of great mechanical inventions there is perhaps no better example of the interdependence of the engineer, the physicist, and the chemist than is evinced in the perfecting of the gas engine. The physicist and the chemist together determine the behaviour of the gaseous fuel, basing their theory on data obtained from the experimental engines constructed by the mechanical engineer, who, guided by their theories, makes his designs and improvements; then, again, from the results of the improvements fresh data are collected and the theory further advanced, and so on till success is reached. But though I have spoken of the physicist, the chemist, and the engineer as separate persons, it more generally occurs that they are rolled into one, or at most two, individuals, and that it is indispensable that each worker should have some considerable knowledge of all the sciences involved to be able to act his part successfully.

Now let us ask, Could not this very valuable invention, the internal-combustion engine, have been introduced in a much shorter time by more favouring circumstances, by some more favourable arrangement of the patent laws, or by legislation to assist the worker attacking so difficult a problem? I think the answer is that a great deal might be done, and I will endeavour to indicate some changes and possible improvements.

The history of this invention brings before our minds two important considerations. Firstly, let us consider the patentable matter involved in the invention of the gas engine, the utilisation for motive-power purposes of the then well-known properties of the explosive energy of gunpowder or of mixtures of gas and oil with air. Are not these obvious inferences to persons of a mechanical turn of mind and who had seen guns fired, or explosions in bottles containing spirits of turpentine when slightly heated and a light applied to the neck? Surely no fundamental patent could have been granted under the existing patent laws for so obvious an application of known forces. Consequently, patent protection was sought in comparative details, details in some cases essential to success which were evolved or invented in the process of working out the invention. In this extended field of operations a slight protection was in some instances obtained. But in answer to the question whether such protection was commensurate with the benefits received by the community at large, there can, I think, be only one reply. Generally, those who did most got nothing, some few received insufficient returns and in very few cases indeed can the return be said to have been adequate. The second important consideration is that of the methods of procedure of the patentees, for it appears that very few of them had studied what had been suggested or done before by others before taking out their own patent. We are also struck by the number of really important advances that have been suggested and have failed to fructify, either from want of funds or other causes, to be forgotten for the time and to be re-invented later on by subsequent workers.

What a waste of time, expense, and disappointment would be avoided if we in England helped the patentee to find out easily what had been done previously, on the lines adopted by the United States and German Patent Offices, who advise the patentee, after the receipt of his provisional specification, of the chief anticipatory patents, dead or alive! And ought we in England to rest content to see our patentees awaiting the report of the United States and German Patent Offices on their foreign equivalent specifications before filing their English patent claims? Ought not our Patent Office to give more facilities and assistance to the patentee?

Before proceeding further to discuss some of the possible improvements for the encouragement and protection of research and invention, I ask you to further consider the position of the inventor—the man anxious to achieve success where others have hitherto failed. To be successful he must be something of an enthusiast; and usually he is a poor man, or a man of moderate means, and dependent on others for financial assistance. Generally the problem to be attacked involves a considerable expenditure of money; some problems require great expenditure before any return can thereby accrue, even under the most favourable circumstances. In the very few cases where the inventor has some means of his own they are generally insufficient to carry him through, and there have unfortunately been many who have lost everything in the attempt. In nearly all cases the inventor has to co-operate with capital: the capitalist may be a sleeping partner, or the capital may be held by a firm or syndicate, the inventor in such cases being a partner—a junior partner—or a member of the staff. The combination may be successful and lasting, but unfortunately the best inventors are often bad men of business. The elements of the combination are often unstable, and the disturbing forces are many and active; especially is this so when the problem to be attacked is one of difficulty, necessitating various and successive schemes involving considerable expenditure, generally many times greater than that foreshadowed at the commencement of the undertaking. Under such circumstances, unless the capitalist or the senior partner or board be in entire sympathy with the inventor or exercise great forbearance, stimulated by the hope of ultimate success and adequate returns, the case becomes hopeless, disruption takes place, and the situation is abandoned. Further, in the majority of cases, after some substantial progress has been made it is found that under the existing patent laws insufficient protection can be secured, and the prospect of a reasonable return for the expenditure becomes doubtful. Under such circumstances the capitalist will generally refuse to proceed further unless the prospect of being first in the field may tempt him to continue.

Very many inventors, as I have said, avoid the expense of searching the patent records to see how far their problem has been attacked by others.

In some cases the cost of a thorough search is very great indeed; sometimes it is greater than the cost of a trial attack on the problem. In the case of young and inexperienced inventors there sometimes exists a disinclination to enter on an expensive search; they prefer to spend their money on the attack itself. There are some, it is true, who have a foolish aversion to take steps to ascertain if others have been before them, and who prefer to remain in ignorance and trust to chance. It will, however, be said that the United States and German Patent Office reports ought to suffice to warn or protect the English patentee; but my own experience has been that such protection is not entirely satisfactory. There is, firstly, a considerable interval before such reports are received, and the life of a patent is short. Then, if the patent is upon an important subject, attracting general attention, the search is vigorous and sometimes overwrought, and the patent unjustly damaged or refused altogether. If, however, the patent is on some subject not attracting general attention, it receives too little attention and is granted without comment.

In some few instances it may be said that ignorance has been a positive advantage, and that if the patentee had realised how much of his patentable work was honeycombed by previous publications and patents, he would have lost heart and given up the task. It is, I think, a case of the exception proving the rule; and the patentee ought, as far as possible, in all cases to know his true position, and make his choice accordingly. The present patent law has some curious anomalies. Let us suppose some inventor has the good fortune to place the keystone in the arch of an invention, to add some finishing touch which makes the whole invention a complete success, and valuable. Then, success having been proved possible, others try to reap the results of his labour and good fortune, and, as often happens, it is discovered after laborious search that someone else first suggested the same keystone in some long-forgotten patent or obscure publication, but for some reason or other the public were none the better for his having done so. What does the law do? It says this is an anticipation, and instead of apportioning to all parties reasonable and equitable shares in the perfected invention, to which no one could object, it says that the patent is injured or perhaps rendered useless by the anticipation, and that its value to everyone concerned is thereby diminished or destroyed, as the case may be, and thrown open to the public. Up till a few years ago, any anticipations, however old, might be cited; but recently the law has been amended, and at present none rank as anticipations which are more than fifty years old.

The perfecting of inventions and their introduction into general use requires capital, as we have seen—sometimes a considerable amount, as in the introduction of the Bessemer process for steel, or the Linotype system of printing—before any commercial success can be realised.

Capital having been found, the next difficulty is in the conservatism of persons and communities who are the buyers of the invention. There is always present in their minds the risk of failure, and its consequent loss and worry to themselves, and in the event of success the advantage, in their estimation, may not be sufficient to counterbalance the risk. In large departments and companies whose management is conducted by officials receiving fixed salaries, acting under non-technical supervision, there is a strong tendency among the officials to leave well alone, the organisation being such that the risk of failure, even though it be remote, more than counterbalances, in their estimation, the advantages that would result in the event of success. Next is the opposition of those who are financially interested in competing trades or older inventions; and if the invention is a labour-saving appliance, then the active opposition of the displaced labour is a serious, though generally only a temporary, barrier.

Fortunately, however, for the community, for research, and for invention, there is always to be found a considerable percentage of persons who, apart from the inventor, are able and willing to risk, and indeed to sacrifice, their personal interests in the cause of progress for the benefit of the community at large; and were it not for such persons the task of the introduction of most inventions would be an impossible one.

There are many problems of the highest importance in physics, engineering, chemistry, geology, and the arts, of which the investigation might probably prove of great benefit to the human race, and of which the probable monetary cost of the attack would be considerable, and of some very great indeed. Let us, then, inquire how the necessary funds could be raised. It is possible in the case of some of the more attractive problems that a group of rich philanthropists might be found, but in most cases it would be impossible to form a company on business lines, under the existing laws of this and other countries, as I shall endeavour to show.

In the case of many of the problems, no patents will give adequate protection; in some cases there is no subject-matter of novelty and importance involved. In other cases the probable duration of the investigation is so long that any initial patents would have expired before a commercial result was reached, and under either of these circumstances there would be no inducement to business men or financiers to undertake the risk.

As an illustration of my meaning I will take two investigations that have doubtless occurred to the minds of most of those present, though many others of greater or less importance might be cited. One is the thorough investigation of the problem of aerial navigation, with or without the assistance of flotation by gas. This problem could undoubtedly be successfully solved by an organised attack of skilled and properly trained engineers and the expenditure of a large sum of money. Assuming the problem solved, and commercially successful, it appears to be impossible under the existing patent laws to secure any adequate monopoly so as to justify the expectation of a reasonable return on the capital expended on the invention. For in view of the multitude of suggestions that have been made and the experiments that have been carried out, the practical solution of the problem would appear to rest on a judicious selection of old ideas by means of exhaustive experiments.

A DEEP BOREHOLE

Another and perhaps more important investigation which has not, as yet, been attacked to any material extent is the exploration of the lower depths of the earth. At present the deepest shaft is, I believe, at the Cape, of a little over one mile in depth, and the deepest bore-hole is one made in Silesia, by the Austrian Government, of about the same depth. What would be found at greater depths is at present a matter for conjecture, founded on the dip and thicknesses of strata observed on or near the surface. Much money and many valuable lives have been devoted to exploration of the polar regions, but there can be no comparison between the scientific interest and the possible material results of such exploration and the one I have chosen for illustration of the inadequate protection afforded by law—namely, a great engineering attack on a problem of geology.

I would ask you to consider the commercial aspect of this engineering geological enterprise, as compared with exploration into new or unknown areas on the surface of the earth.

An exploring expedition into a new country has before it generally the probability of the acquisition of territorial and mineral rights or possessions bringing material gain to the undertakers. The rights of such enterprises are well known, and capital can be obtained with or without national support, as the case may be. On the other hand, the explorer into the depths of the earth has no rights or monopolies beyond the mineral rights of the land he has purchased over his boring; further, it is improbable that he can obtain any patent of substantial value for his methods of boring to great depths. To succeed in the undertaking a great expenditure of money must be incurred, an expenditure far greater than that of an exploring expedition, and analogous to that of a military expedition or a small invading army, and to raise this sum the pioneers have practically no security to offer. For if they succeed in finding rich deposits of precious minerals in greater abundance, or succeed in making some geological discovery associated with deep borings, they gain no exclusive title to these under existing laws. Any other person or syndicate, acting upon the experience gained, could sink other shafts in other places or countries, and, benefiting by the experience gained by the pioneers, could probably carry out the work more advantageously, and thus depreciate the first undertaking or render it valueless, as has often occurred before.

Let us consider more closely some of the essential features of sinking a shaft to a great depth, for I think it will be seen that it presents no unsurmountable difficulties beyond those incidental to an enterprise of considerable magnitude involving the ordinary methods of procedure and the ordinary methods adopted by mining engineers. That there would be some departures from ordinary practice on account of the great depth it is true, but these are more of the character of detail. On the design of this boring I have consulted Mr John Bell Simpson, the eminent authority on mining in the North of England. The shaft would be sunk in a locality to avoid as far as possible water-bearing strata and the necessity of pumping. It would be of a size usual in ordinary mines or coal-pits. The exact position of such a shaft would require some consideration as to whether it should commence in the primary or secondary strata. It would be sunk in stages, each of about half a mile in depth, and at each stage there would be placed the hauling and other machinery, to be worked electrically, for dealing with each stage. The depth of each stage would be restricted to half a mile in order to avoid a disproportionate cost in the hauling machinery and the weight of rope, as well as increased cost in the cooling arrangements arising from excessive hydraulic pressures. At each second or third mile in depth there would be air locks to prevent the air pressure from becoming excessive owing to the weight of the superincumbent air, which at from two to three miles would reach about double the atmospheric pressure at the surface. A greater rise of pressure than this would be objectionable for two reasons—firstly, from the inconvenience to the workmen; secondly, from the rise of temperature due to the adiabatic compression of the circulating air for ventilating purposes. The air pressure immediately above each air lock would thus reach to about two atmospheres, and beneath to one atmosphere. In order to carry on the transfer of air through the air locks for ventilating purposes pumps coupled to air engines would be provided, the energy to work the pumps being obtained from electro-motors. To maintain the shaft at a reasonable temperature at the greater depth powerful means of carrying the heat to the surface would be provided.

The most suitable arrangement for cooling would probably consist of large steel pipes, an upcast and a downcast pipe, connected at the top and bottom of each half-mile section in a closed ring. This ring would be filled with brine, which by natural circulation would form a powerful carrier of heat; but the circulation, assisted by electrically driven centrifugal pumps, would be capable of carrying an enormous quantity of heat upwards to the surface. At each half-mile stage there would be a transfer of the heat from the ring below to the ring above by means of an apparatus similar in construction to a feed-water heater, or to a regenerator constructed of small steel tubes, through which the brine in the ring above would circulate, and around the outside the brine in the ring below could also circulate, the heat being transmitted through the metal of the tubes from brine ring to brine ring.

We have now presented to us two alternative arrangements for cooling. One arrangement would be to cool the brine to a very low temperature in the top ring at the mouth of the shaft by refrigerating machinery, so as to provide a sufficient gradation of temperature in the whole brine system, to ensure the necessary flow of heat upwards from brine ring to brine ring, and overcome all the resistances of heat transfer, and so maintain the lowest ring at the temperature necessary for effectual cooling of the lowest section of the shaft. But a better arrangement would be to place powerful refrigerating machinery at certain of the lower stages, the function of this machinery being to extract heat from the ring below and deliver it to the ring above. This latter method would increase to a very great extent the heat-carrying power of the system, which in the first arrangement is limited by the freezing temperature of brine in the descending column and the highest temperature admissible in the ascending brine column. The amount of heat conducted inwards through the rock wall and requiring to be absorbed and transferred to the surface depends on the temperature and conductibility of the strata. But there is no doubt that the methods I have indicated would be capable of maintaining a moderate temperature in the shaft to depths of twelve miles.

During the process of sinking at the greater depths the shaft bottom would require the application of a special cooling process in advance of the sinkers, similar to the Belgian freezing system of M. Poesche used for sinking through water-bearing strata and quicksands, and now in general use. It consists in driving a number of bore-holes in a circle outside the perimeter of the shaft to be sunk; through these bore-holes very cold brine is circulated, thus freezing the rocks and quicksands and the water therein, and when this process is completed the sinking of the shaft is easily accomplished.

In our case this process would be maintained not only on the shaft bottom, but also for some time on the newly pierced shaft sides, until the surrounding rock had been cooled for some distance from the face.

As to the cost, rate of boring, and normal temperature of the rock, an approximate estimate has been made, based on the experience gained on the Rand, but including the extra costs for air locks and cooling:

	Cost £	Time in years	Temperature of rock
For 2 miles depth from the surface	500,000	10	122° F.
4 ,, ,, ,,	1,100,000	25	152°
6 ,, ,, ,,	1,800,000	40	182°
8 ,, ,, ,,	2,700,000	55	212°
10 ,, ,, ,,	3,700,000	70	242°
12 ,, ,, ,,	5,000,000	85	272°

I hope I have succeeded in showing in the short time at our disposal that an exploration to great depths is not an impossible undertaking. But my main object in discussing the enterprise at some length has been to show that a pioneer company would not acquire any subsequent monopoly of similar works under the existing patent laws or the laws of any country.

In the scheme as I have described it, there appears to be nothing that could be patented; but let us suppose that some good patent could have been found that was absolutely essential to the success of the undertaking, it would certainly have expired before the pioneer company could have reaped any substantial return, and probably before the first enterprise had been completed. It follows therefore that at the present time there is no adequate protection, or indeed any protection at all, for the promoters of many great and important pioneer enterprises, some of which might prove of immense benefit to mankind.

Let us ask what change in the laws would place great pioneer research works on a sound financial basis. A Government grant, except for very special purposes, seems to be out of the question, seeing that the benefits to be derived are generally not confined to any one country. An extension of the life of patents, which is now from fourteen to sixteen years in different countries, would be undoubtedly a step in the right direction. It would be of great benefit generally if some scale of duration of patents could be fixed internationally, the scale being fixed according to the subject-matter, the difficulty of the attack, and the past history of the subject, but more especially in view of the utility of the invention.

One of the chief objections raised by the Privy Council against the extension of patents in this country has rightly been that undue prolongation is unfair to the British public, seeing that abroad no prolongations are granted. Therefore, if the duration of patents for important matters is to be extended at home, it must also be extended abroad. In other words, such prolongations, to be effective, should necessarily extend to other countries. They should be international, and concurrent in all the countries interested.

One possible solution of this difficult question would be to place such matters under the jurisdiction of a Central International Committee, who would have the apportionment of the life and privileges of patents and of the extension or curtailment of their duration, according to their handling by the owners. I would ask, Why has a patent a life of only fourteen to sixteen years, while copyright is for forty-two years? Why has a pioneer company making a railway under Act of Parliament generally rights for ever unless it abuses its privileges, or the requirements of the district necessitate the construction of competing lines, while a patent has in comparison a life of infinite shortness?

I might also cite gas companies, electrical supply companies, under Act of Parliament, or provisional orders of forty-two years' duration; and this reminds us of the fact that until the term of life for electric supply companies had been extended from twenty-one years to forty-two years by the bill of 1884, it was impossible to find capital for such undertakings.

Now, it may be urged that the grant of a patent is a different thing from the grant of power to a railway company, a gas or electric supply company. But the object of this Address has been to show that a patent, to be fair to the patentee, ought in many cases to be analogous to an Act of Parliament or a provisional order. Would it not place matters in a fairer position, especially in the case of expensive and lengthy researches, to grant to those who pledge themselves to spend a suitable and minimum sum within a stated period on the research a reasonable and fair monopoly, so that such person or syndicate might in the event of success be in the position to reap a reasonable return for their expenditure and risk?

Some such measure would unquestionably give an immense stimulus to research and invention by enabling capital to be raised and works started on commercial lines in fields of great promise at present almost untouched.

I pass over the disadvantages to the British inventor of the hostile patent tariffs of Continental nations and of the protective patent laws of some of the British dependencies, disadvantages greater than those imposed by protective tariffs on the ordinary British manufacturer.

There is, however, another aspect of the question to which I would briefly allude: it is the great benefits that the world at large has derived from the work of inventors in the past.

Think of the multitude and power of the great steam engines and gas engines that drive our factories, and pump the water out of our mines, and supply our cities with water, light, and power; of the great steamships scattered over the ocean and the locomotives on the railways.

Think of the billions of tons of steel that have been made by the Bessemer, Siemens-Martin, and Thomas-Gilchrist processes, and of the great superiority and less cost of the material over the puddled iron which it superseded.

Think of the vast work performed by the electric telegraphs and telephones, and we must not fail to include the great chemical and metallurgical processes carried on all over the world, besides the countless other inventions and labour-saving appliances.

Can we form any idea of the commercial value of all these gigantic tools that past inventors have left as a heritage to the human race, and can we venture to place any order of magnitude on so vast a sum?

If we take as our unit of value the whole of the money spent on all inventions, both successful and unsuccessful, I think we shall be much below the mark if we assume that the value of the benefits has on the average exceeded by ten-thousandfold the money spent on making and introducing the inventions.

If this is so, let us see what it means. It means that for every unit of capital spent by the inventors and their friends on invention they have in some cases received nothing back. In some cases they have just got their capital back, in some cases two or threefold, occasionally tenfold, very rarely a hundredfold. Whereas the world at large has received a present of ten-thousandfold greater value than all the money spent and misspent by the small band of past inventors.

In conclusion, let us hope that the inventor will in the future receive more encouragement and support, that the patent laws will be further modified and extended, that the people at large will consider these matters more closely and recognise that they are of first importance to their progress and welfare, and that in the future it may be easier, nay in some cases possible, to carry on many great researches into the secrets of Nature.

THE STEAM TURBINE
(With George Gerald Stoney, B.A., M.Inst.C.E.)

Read at the Institute of Civil Engineers, December 5th, 1905

The manufacture of the Parsons steam turbine was begun systematically in 1885, the design and constructive details being gradually improved. By the year 1889, turbines aggregating about 4000 horse-power had been put to work; and further improvements in the formation and curvature of the blades, together with the arrangement of the turbine for use with a condenser, resulted in such a reduction of the steam consumption, that in 1892, as will be evident from the following table, the new motor had attained to the position of a competitor with the best compound reciprocating engines, and the prospect of its use for many purposes, including the propulsion of ships, began to come into view.

Date	Place	Power (kilowatts)	Steam per kilowatt hour (lb.)	Vacuum (inches)	Super-heat (°F.)	Steam pressure per sq. in. (lb.)
1887	—	75	50	—	—	120
1892	—	100	27·00	27	50	—
1898	—	1250	18·81	28	180	130
1901	—	1000	17·30	27	198	150
1902	Frankfort-on-Main	3000	14·74	27	235	138
1904	Carville station, Newcastle Electric Supply Co.	4000	15·40	28·7 (Bar. 30)	150	200

Fig. 1, which is based upon a large number of tests, shows the steam consumption of different-sized plants per kilowatt hour, without superheating; and it is interesting to note that the curve shows that the efficiency is increasing steadily as the size grows larger.

Curves showing the variation of steam consumption of a turbine with different degrees of vacuum are given for a 300 kilowatt plant in Fig. 2, and for a 1500 kilowatt plant in Fig. 3. It is interesting to note from Fig. 3 that in a turbine for 1500 kilowatts at two-thirds of the normal output, the effect of an increase of 1 inch of vacuum at 26 inches is to diminish the consumption by about 4 per cent.; at 27 inches by $4\frac{1}{2}$ per cent.; at 28 inches by $5\frac{1}{2}$ per cent.; and between 28 inches and 29 inches by 6–7 per cent. A good vacuum is thus seen to be much more essential for the steam turbine than for the reciprocating engine, as might be anticipated on account of the greater range of expansion dealt with in the former. The effect of difference

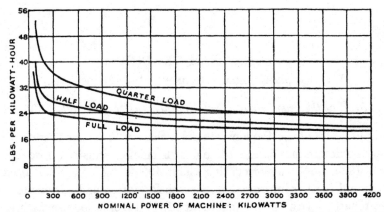

Fig. 1. Steam pressure 150 lb. per square inch. Vacuum at full load 27 inches, at half load 27½ inches, at quarter load 28 inches. Barometer 30 inches.

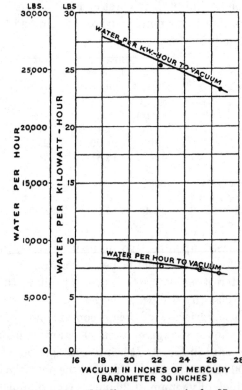

Fig. 2. Steam pressure 155 lb. per square inch. No superheat.
Mean speed 3000 revs. per minute.

of boiler pressure, however, is relatively smaller in turbines than in reciprocating engines, and it is questionable whether in most cases the saving in coal by the adoption of boiler pressures higher than 150–200 lb. per

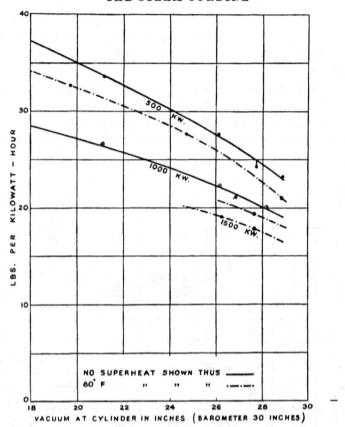

Fig. 3. Steam pressure 120 lb. Speed 1500 revs. per minute

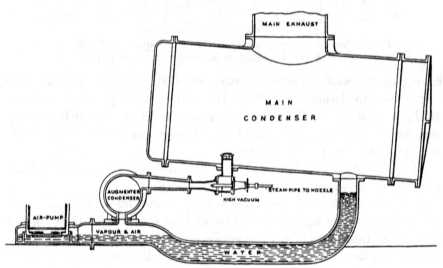

Fig. 4. Vacuum Augmentor.

square inch is sufficient to justify the increase. Superheating of the steam has a very marked effect in diminishing the steam consumption, for it is found that every 10° F. of superheat reduces the steam consumption by about 1 per cent.

The question of condensing plant is such an important factor with turbines, that some remarks on this subject seem necessary. Speaking generally, with well-designed air and circulating pumps, and with a condenser

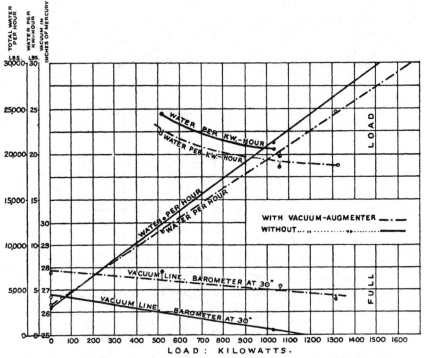

Fig. 5. Mean stop-valve pressure 139 lb. per square inch. Mean superheat 115° F. Mean speed 1600 revs. per minute.

having an allowance of about 1 square foot of surface per indicated horse-power and circulating thirty times the feed at a temperature of about 70° F., a vacuum of 26–27 inches can usually be maintained with ordinary supervision. In order to obtain a higher vacuum the following conditions and arrangements must be observed:

(*a*) The condenser surface must be increased to about 1½ square feet per indicated horse-power.

(*b*) It is necessary to increase the volume of the circulating water, and so to modify the design of the condenser that the velocity of the water through the tubes shall be 4–7 feet per second, so as to secure some measure of turbulent flow and consequently better absorption of heat.

(*c*) The tubes of the condenser should be so spaced as to allow an easy

Data from 1500 kilowatt turbo-alternator, with and without vacuum augmenter

Pressure above atmosphere (lb. per sq. in.)	Superheat (°F.)	Vacuum		Cooling-water temperatures			Speed (revs. per min.)	Load (kw.)	Steam used including 450 lb. per hour in vacuum augmenter	
		At turbine cylinder (in.)	At condenser (in.)	Inlet (°F.)	Outlet (°F.)	Air-pump discharge (°F.)			(lb. per hour)	(lb. per kw.)
				With vacuum augmenter.						
113·6	108·3	26·69	26·95	67·0	91·0	104·0	1445	1316·5	24,732·0	18·76
111·6	156·4	27·12	27·4	64·5	85·0	99·0	1500	1061·6	19,830·0	18·66
141·0	113·0	27·72	27·85	60·0	72·5	83·0	1500	512·7	11,425·0	22·3
154·0	47·5	27·72	27·92	59·5	62·5	92·6	1500	—	3,127·7	—
				Without vacuum augmenter.						
115·6	143·0	25·18	25·32	63·0	84·0	92·0	1500	1029·3	21,264·0	20·7
137·0	119·0	25·97	26·12	61·0	72·0	75·6	1500	534·02	12,820·0	24·02
150·3	72·4	26·62	26·82	58·0	61·5	64·0	1500	—	2,957·4	—

flow of steam among them; and the authors have also found it desirable to submerge the lower tubes in the condensed water before it goes to the air pump. This is done by providing, in the bottom of the condenser, a weir which holds up the condensed water so as to cover the bottom two or three rows of tubes. By this arrangement the condensed water is thoroughly cooled before it reaches the air pump, and a lower air pump temperature is obtained.

(d) It is generally necessary to use a larger air pump; but the increase of power required for this and for the larger circulating pump (when the circulating water has not to be lifted more than 15–20 feet, and the return main is water-sealed) will not exceed $1\frac{1}{4}$ per cent. of the total power developed.

With the foregoing arrangements it is possible to maintain a vacuum of $27\frac{1}{2}$–28 inches, and thus to effect a saving of 5 per cent. on the coal consumed.

Because of the importance to the steam turbine of a high vacuum the authors have been led to the design of a new apparatus to assist the ordinary air pump and condenser. This apparatus, which is shown diagrammatically in Fig. 4, consists merely of a steam jet, placed in a contracted pipe between the condenser and the air pump, which draws the air and vapour from the condenser and reduces the vapour density therein to about one-third. The mixture of air and vapour is compressed in the contracted pipe to about one-half of its bulk, and is delivered to the air pump through a small auxiliary condenser (having about 2–3 per cent. of the cooling surface of the main condenser), which cools the air and partially condenses the vapour before it enters the air pump. The water of condensation in the main condenser gravitates directly into the air pump suction, by a separate connecting pipe, which is water sealed. The consumption of steam in the jet has recently been reduced to about 1–$1\frac{1}{2}$ per cent. of that being dealt with at normal load in the main condenser; and the observed total net reduction of steam consumption in the turbine is about 8 per cent. at full load, the condenser, the volume of circulating water, and the velocity of the air pump, being the same. Fig. 5 shows the results of this vacuum augmenter on a 1500 kilowatt plant between no load and full load; the consumption of steam includes that of the jet, which at the time of the trial was 450 lb. per hour. The Table gives the temperatures, etc., in different parts of the system. In removing nearly the whole of the air from the condenser, the apparatus also brings the steam into more intimate contact with the surface of the tubes, and it is found in practice that by its use the vacuum can be raised from 26 inches to $27\frac{1}{2}$ or 28 inches without any alteration to the plant, except in some cases an increase in the volume of the circulating water.

PLATE IV

The engines of H.M. Destroyer *Viper*

THE STEAM TURBINE ON LAND AND AT SEA

Royal Institution, Friday, May 4th, 1906

It was with some diffidence that I accepted the subject of Steam Turbines on Land and at Sea for this evening's lecture, for since I had the privilege of dealing with this subject six years ago in this room, there seemed to me to be very little new to add, either from a scientific or a practical point of view, which had not then been to some extent considered. However, after consideration, there seemed to be a hope that an account of some further developments during the last six years on land and on sea, and a more extended description of the mechanics of the turbine and its applications, might prove of some interest, in view of the more general adoption of the turbine principle for the generation of electricity, for the propulsion of vessels, and for driving air compressors, fans, and pumps.

Six years ago there were 75,000 horse-power of turbines on land, and 25,000 on sea. At the present time there are more than two million horse-power at work on land, and 800,000 horse-power at work or building for use at sea.

There are at present afloat, equipped with turbines:

3 Pleasure steamers.	6 Yachts.
9 Cross-channel steamers.	3 Destroyers.
5 Ocean-going vessels.	2 Cruisers.
3 Atlantic liners.	

Yet it cannot be said that the turbine engine is superseding the reciprocating engine generally, although this is undoubtedly to some extent the case in certain fields of work.

On land, the chief application of the turbine is found in large electrical generating stations, and its adoption in preference to the piston engine, in its most perfect development of compound, triple, or quadruple expansion engine, is becoming general in this field of work.

At sea, its use is commencing to extend for all the larger and faster class of ships; for cross-channel steamers it has found great favour, and for Atlantic liners and ships of war it is being used to a more and more considerable extent, and this tendency is not confined alone to England, but is shown also on the Continent, and in the United States and Japan. It will give a clearer idea of the subject if we first of all examine more closely the characteristics of the steam turbine, and generally how it works.

All turbines derive their power from the impact of the steam, or, more correctly speaking, from the momentum of the steam, flowing through them, just as a windmill receives its power from the wind.

There are three principal types of turbines now in general use, as well as some which may be described as admixtures of these three classes. They differ essentially in some respects, more particularly in their methods of extracting the power from the steam.

The first to receive commercial application, 1884, was the compound or multiple expansion steam turbine; the second was the De Laval or single-bucket wheel, in 1888, driven by the expanding steam jet; and lastly, the Curtis turbine, in 1896, which comprises some of the principal features of the others combined with a sinuous treatment of the steam.

In the compound turbine, the steam is caused to flow through a series of many turbine elements of gradually increasing size, graduated so as to allow of the expansion of steam in small increments of volume at each element, these increments of volume corresponding to the fall of pressure necessary to cause the steam to flow through each element. Each element consists of a row of guide blades and a row of moving blades. The guide blades are attached in circumferential rows to the case and project inwardly, and the moving blades are attached in rows to a drum and project outwardly. The ends of the blades throughout the turbine nearly touch the drum and case respectively.

To form some idea of the forces at work in a turbine we should consider, with approximate accuracy, that the steam flows through the turbine with a force about ten times as great as that of the strongest hurricane; and though the force acting on each blade is small, perhaps only a few ounces, or in the largest only a few pounds, yet in the aggregate the force is great and can propel large ships or drive large dynamos.

The important factors upon which the proportions of the turbine are based are the pressures, velocities and percentages of moisture in the steam, as it gradually expands from turbine row to turbine row.

The blades of the turbine are made of rolled and drawn brass, well shaped, and polished so as to reduce the frictional losses in the steam to a minimum. The steam enters all round the shaft and first traverses the shortest blades on the smallest drum, then through larger and larger blades set on larger and larger drums, and so on till as it leaves the last blades it is expanded about 100-fold in volume. At the opposite end to the blade drums are the balance pistons, or dummy drums, which serve to balance the end pressure of the steam, and are kept steam tight with the casing by packing grooves on the dummy drums which rotate in close proximity to corresponding but stationary brass rings keyed into the case.

In land turbines, for driving dynamos or other fast moving machinery, no end pressure on the shaft is required, nor is it permissible because of the mechanical difficulties met with in thrust-bearings carrying heavy end pressure and rotating at high speed, and therefore balance pistons are provided, which, while being practically steam tight, serve to balance all end pressure arising from the steam acting upon the rotating barrels and vanes.

In marine turbines, on the other hand, the dummy drums are so proportioned as to leave an unbalanced end pressure, which counteracts and balances the thrust of the propeller, thus relieving the thrust bearing from pressure.

The bearings of the engine, it will be seen, have only to support the weight of the rotating part of the engine; this is comparatively small, and as continuous lubrication is provided by an oil pump which circulates the oil continuously through the journals round and round, there is practically no wear, even after years of continuous work; and the maintenance of the shaft in a truly central position relatively to the casing, which is of great importance, is easily maintained in practice.

Before proceeding further with the examination of the compound steam turbine, let us consider the De Laval steam turbine introduced by Dr De Laval of Stockholm in 1888.

The De Laval turbine

In this turbine the steam at full pressure issues from a diverging conical jet, so formed and proportioned that the steam after passing through the neck of the jet enters a gradually divergent passage of increasing cross section, in which it expands; the result being that nearly the whole available energy in the steam is utilised in imparting to it a very high velocity, reaching, with 100 lb. boiler pressure and a good vacuum, as much as 4200 feet per second. The discovery of this property of the expanding jet is due chiefly to Dr De Laval.

This rapidly moving column of expanded steam is directed against cupped steel buckets on the periphery of a wheel made of the strongest steel, the wheel being shaped so as to permit of the highest peripheral velocity consistent with safety, which may be from 800 to 1200 feet per second; the steam, by striking the cups and reacting, partly by velocity of flow and partly by elastic gaseous rebound from the concave surface of the cups, leaves the wheel with a considerable backward velocity, and to obtain the highest efficiency it is necessary to reduce this backward velocity by increasing the velocity of the wheel to the uttermost. The strongest materials, however, do not permit of a close approach to the speed necessary for the maximum efficiency; yet in this turbine, owing to the comparative absence of losses, which are present to some extent in the other types (and which we will consider presently), the efficiency of this turbine compares favourably for moderate and small powers.

In this beautiful construction, developed with mechanical skill and guided by an intimate acquaintance with the properties of steam and materials, there are many minor features of interest. Among them may be mentioned the elastic shaft, to permit of the rotation of the turbine wheel about its dynamic axis. A device, consisting of frictional damping washers, which had the same purpose as this elastic shaft, was used in 1885 in the early development of the compound steam turbine. It was superseded in 1892 by the damping effect of thin films of oil between several concentric loosely fitting tubes surrounding the bearings.

The De Laval turbine has for many years been extensively used on the Continent and in this country, in sizes up to about 400 horse-power. Its chief use has been for the driving of dynamos, pumps, fans, and motive power generally; but, owing to its very high angular speed, it is necessary in most cases to use gearing, except when driving very fast running centrifugal pumps and fans.

The gearing is of steel, and it is accurately cut with very fine spiral teeth, and it works satisfactorily even at a speed of 30,000 revolutions per minute.

THE CURTIS TURBINE

Let us now consider the Curtis turbine. It ranks in a class by itself, because it comprises the principle of the sinuous treatment of expanded steam first put into extended commercial use by Mr Curtis under the auspices of the General Electric Company of America.

This sinuous treatment of the steam consists in giving to it a high initial velocity by passing it through a jet of the De Laval type, or a group of such jets; it then impinges on a ring of bucket blades like those used by De Laval, and after leaving the first row of such blades it is caught by a ring or a sector of stationary bucket blades set in the reverse direction, and by them its direction is changed into that of the next succeeding row of moving blades (there may be three rows of moving blades in all and two sectors of fixed blades); and the height of each succeeding row is increased, to allow a greater area for the steam as it flags in velocity after each rebound between the moving and fixed blades.

The object of this treatment is to transfer a large percentage of the kinetic energy of the rapidly moving steam to the moving blades and wheel, without the necessity of very high peripheral speeds of blades, such as are necessary with the single-wheel type. As regards, however, "multiple series action", the principle resembles the compound turbine.

The expansion process in nozzles, and subsequent sinuous treatment of the steam, is repeated several times by four or more similar wheels on the same axis, but in separate steam-tight chambers, until the steam is fully expanded.

If there are four such operations, the velocity of outflow from the nozzles will be about 2000 feet per second, and the peripheral velocity of wheel about 400 feet per second; at each operation the steam is expanded through one-fourth of the whole range, and at each it is brought to rest before flowing to the next chamber through the jets.

A great many other varieties of the turbine have been proposed, and some have received a limited application. The Rateau, the Riedler-Stumpf, the Zoelly, the Escher Wyss, and many others, might be mentioned as varieties of the three fundamental turbines we have considered; indeed in some cases the variation would appear to have been only a retrograde step, and represents some discarded form tried by one of the originators of the three fundamental types.

As far as we can gather from the history of the steam turbine, it may be said broadly that all the chief features at present in use in turbines have been suggested or described in the rough by experimenters long ago in the hundred and more patents prior to 1880.

For instance, Hero of Alexandria, 130 B.C., made a reaction wheel.

William Gilmore first suggested the compound steam turbine in 1837.

Matthew Heath first enunciated the principle of the diverging conical jet in 1838.

James Pilbrow in 1842 used cupped buckets, and suggested a sinuous treatment of the steam.

Robert Wilson developed the compound steam turbine to a considerable extent in 1848.

It would take too long to trace the initiation of each idea, but we may say, in the light of recent experience, that most, if not all, the designs showed a want of knowledge of the properties of steam and materials, and could not have given a satisfactory performance*.

THE PARSONS TURBINE

Let us again recur to the compound turbine, and look more closely into the principles of its working, and more particularly consider the course of the steam in its passage through the vanes or blades of the engine.

Viewing the turbine as a whole, we see that the steam passes through the forest of fixed and moving blades just as water flows from a lake of higher level through a series of rapids and intervening pools to a lake of lower level. The boiler corresponding to the lake of higher level and the condenser to that of lower level.

In the flow through the turbine the steam is repeatedly gathering a little velocity from the small falls of pressure, which is as soon checked and its

* [It is rather strange that Parsons nowhere refers to Richard Boyman Boyman whose patent of 1860 is a remarkable document. Ed.]

energy transferred to the blades, over and over again; 50 to 100 times is this repeated before it is fully expanded and escapes into the condenser.

The number of blades in a steam turbine is very great; in a 2000 horse-power engine it may be from 20,000 to 50,000 and the surface speed of the several barrels of the turbine will be from 150 to 300 feet per second. In such an engine it is arranged that the lineal velocity of the blades shall approximate to one-half that of the tangential component of the steam issuing from the guide blades. The blades are curved, with thickened backs, and are smooth; the steam therefore flows around them, and past them, without much loss by shock or eddy current or frictional loss. The proportions of turbines as regards diameter, height of blade, and blade openings, are calculated so that, under average working conditions, the correct expansion of the steam shall be attained, and the fall in pressure and velocity of steam at each turbine of the series shall be such as to secure for it the highest efficiency.

When a turbine is tested the pressures at many points along the barrel are recorded, and the calculated pressures confirmed and verified by experiment, and these are usually in close accord. As the result of data accumulated from experiments on many turbines, the probable horse-power that will be obtained from a given design of turbine can be predicted with as much accuracy as in the case of the reciprocating engine. The best results that have been obtained from large turbines show that about 70 per cent. of the available energy in the steam is converted into brake horse-power; and where, we may inquire, has the other 30 per cent. gone to?

The chief losses of efficiency in all steam turbines are due to three principal causes: first, to skin friction of the steam coursing at high temperature through the small openings between the blades; secondly, to unavoidable leakages; and, thirdly, to eddy-current losses arising from insufficient blade velocity and errors of workmanship.

The first of these losses, the friction of the steam, is reduced by super-heating, and thus partially removing the fluid frictional loss arising from the drops of condensed water mingled with the steam. In some cases this gain in efficiency is worth the extra cost of the superheater, but, unless intermediate superheaters are used, initial superheat cannot be raised high enough to maintain dryness throughout the major part of expansion, without destroying the turbine. Moderate initial superheat, however, is generally used with some gain in economy, which in the compound turbine amounts to 1 per cent. for every 10° F. of superheat. The second loss, which is from leakage, is present in the compound and the sinuous types but not in the De Laval type. The amount of this loss decreases as the size of the engine increases. It is also chiefly consequent on the coefficient of expansion of metals, which is a bugbear to the turbine designer. If a metal with a much smaller coefficient of expansion than steel and iron could be obtained at a reasonable price and of suitable qualities for the construction of turbine cases, drums, and shafts, a considerable increase of economy could be obtained, as it would allow of smaller working clearances and less leakage. The third loss, from insufficient blade velocity, is not present to a material extent in the larger compound or sinuous course turbines, but is present, as already explained, to a considerable extent in the single-wheel type.

Reviewing more closely the motion of the steam through the blades of a compound turbine, we see that the portion of its course during which it is travelling at relatively high velocity, and in close proximity to the blades, is short in comparison with the total length of its travel within the turbine. The passage ways between the blades constitute virtually jets of rectangular cross section, but having easy curves, and the frictional losses are consequently small. After leaving the blades, the steam traverses the intervening space in the form of an annular cylinder with a spiral motion, the angle of pitch being about 30° to a plane normal the axis; and, as the succeeding blades are moving in a similar direction to this flow, we see that the velocity with which the steam is cut by their frontal edges is much less—in fact, less than one-half the velocity at which the steam has issued from the previous blades. From this we see how small is the loss due to the cutting of the steam by the frontal edges in the compound turbine, and also how small is the velocity with which drops of water strike the metal of the blades.

This is an important feature.

It has been shown by experiment that if drops of pure water, arising from the condensation of expanding steam, impinge on brass at a greater velocity than about 500 feet per second there results a slow wearing away of the metal. It is very slow, and would require about ten years to erode the surface to a depth of 1/32 inch. In the compound turbine the striking velocity is much below this figure, and the preservation of the form and smoothness of surface of blades has been found to be practically indefinite.

It appears that the erosive power of drops of pure water moving at high velocity increases rapidly with the velocity, it may probably be as the square. Experiment has shown that if saturated steam at 100 lb. pressure be allowed to flow through a divergent jet into a good vacuum, attaining a velocity of about 4500 feet per second, and allowed to impinge on a stationary brass blade, the blade will be cut through in a few hours, and the hardest steel will be slowly eroded. The action seems to be the result of the intense local pressure from the bombardment of the drops, which may exceed 100 tons.

Owing to the receding velocity of the blades from the blast, and consequently the reduced striking velocity, the erosion of the blades in impact turbines is much reduced, and in compound turbines there is complete immunity from such erosion.

It may be asked, how is it that the steam turbine in the larger sizes is more economical in steam per horse-power developed than the best triple or quadruple expansion reciprocating engine? The reason is, that all large steam turbines are able to take full advantage of the whole expansive energy of the steam, even when expanding to the very attenuated vapour densities produced by the best condensers. It is indeed easy to construct the low-pressure portion of the turbine to deal effectively with the very attenuated vapour, whereas the reciprocating engine, from its nature, can only take full advantage of about two-thirds of the whole range of expansion, and is unable to deal usefully with very low vapour densities—the low-pressure cylinders cannot (because of structural difficulties) be made large enough, and the last part of the expansion has to be allowed to run to waste.

The growth in size of the turbine is perhaps interesting. The first practical steam turbine, constructed in 1884, was of 10 horse-power. By 1892 the largest size for driving dynamos had reached 200 horse-power. It has been continuously increasing, and has now reached 12,000 horse-power in one unit driving one alternating dynamo.

MARINE STEAM TURBINES

The application of the turbine to the propulsion of vessels involved some interesting problems. The most important was, how slow could a turbine be made to rotate consistently with the maintenance of its efficiency in steam consumption, and at the same time be of moderate weight and cost?

In the same problem naturally arose the question of how fast could a screw propeller be made to revolve when propelling a vessel of a given size and at a given speed—in other words, when delivering a given propulsive horse-power at a given speed? The first question as to designing a low-speed turbine was solved in 1894 to 1896, by the aid of the accumulation of accurate data from experiments on land turbines; and the modification arrived at in the turbine has been chiefly directed to the splitting of it up into two or three or more turbines in series on the steam, and each working a separate shaft. This splitting up of the turbine results in a twofold advantage. It makes the turbine (which otherwise would be very long) much shorter, and because of being shorter with finer clearances less loss by leakage results, and the whole engine is lightened. A secondary gain, resulting from the division of the power over several separate shafts, arises from the fact that smaller propellers may be used, making higher speeds of rotation admissible, which again acts in lightening and improving the economy of the turbines.

PLATE V

Parsons Vacuum Tank for propeller testing, 1910

The second question, that of the propeller, was much more difficult. It was not simply the problem of designing a screw with a moderate slip ratio and a moderate loss by skin friction of the blades in the water, but it was complicated by cavitation, or the hollowing out of the water and the production of vacuous cavities caused by the force of the blades tearing through the water, a phenomenon first noticed by Sir John Thornycroft and Mr Sidney Barnaby in 1893, and by them named cavitation. This apparatus shows the phenomenon.

[A small tank was shown, with a model of the screw of a cross-Channel boat or of an Atlantic turbine liner. It was pointed out that it was very difficult to make the screw cavitate, because it was especially designed not to cavitate; it was, however, made to do so in the tank by removing the atmospheric pressure from the surface of the water above the propeller by an air pump. The removal of the atmospheric pressure, which helped to keep the water solid, enabled cavitation to be induced at a much lower speed of revolution. In the tank there was a head of about $1\frac{1}{2}$ inches of water above the topmost blades. If the tank had not been exhausted there would have been a head equivalent to 32 feet, plus $1\frac{1}{2}$ inches, plus capillary forces, tending to keep the water solid. Therefore, instead of 1500 revolutions (the speed of the propeller when serious cavitation was induced) a speed of at least 20,000 revolutions would have been required (because forces that induce cavitation vary as the square of the surface speeds of the blades).]

Serious cavitation causes an inordinate loss of power, chiefly because it disturbs the steam lines around the propeller blades, and it was proved by this experiment how easy it is to put too much work on a screw. There is a limiting thrust that it will bear, and if we exceed this thrust it will, so to speak, more or less strip its thread in the water and its efficiency will rapidly fall. The solution of the problem, as regards the screw propeller, has therefore resulted in a modification of the proportions of the ordinary propeller, and has lain in the direction of smaller diameters, wider blades, and a slightly finer pitch ratio, which three slight changes combined have led towards higher angular speeds of the propeller without material loss of efficiency.

Let us now turn our attention to the economic results of the steam turbine. In the case of large engines and dynamos that are coming generally into use, for the generation of electricity in this and other countries, of a horse-power of 1000 to 12,000 and upwards, the steam turbine with its accompanying dynamo is found to be cheaper in first cost, running expenses, and fuel, than the reciprocating engine and its slow-speed dynamo; and so much is this the case that it seems possible to generate electricity in colliery districts almost, if not quite, as cheaply for electro-chemical purposes as it can be produced at Niagara and some other large centres of water power.

The chief items in which saving has resulted as compared with the reciprocating engine are: the total capital cost of the station is reduced by from 25 to 40 per cent.; the reduction in the cost of fuel and boilers is between 10 and 30 per cent., and the consumption of oil is reduced to one-sixth, while the engine-room staff is reduced by 25 to 50 per cent.

As to the economic results of turbine vessels compared with vessels propelled with piston engines, reliable statistics are available.

In 1897, the *Turbinia* was found to have an economy in steam per horse-power developed, equal to, if not superior to, that of similar vessels propelled by reciprocating engines; and later, in 1903, she was again tried with modified propellers as now generally used which gave a further increase of efficiency of about 10 per cent. over the 1897 trials.

In 1902, the first turbine passenger boat, *King Edward*, on the Clyde, was found to consume about 15 per cent. less coal than a similar vessel propelled by triple expansion engines and twin screws.

The trials of the third-class cruiser *Amethyst*, in 1904, and of her sister vessel the *Topaz*, propelled by triple-expansion engines and twin screws, showed that, at a speed of 11 knots, the consumption of steam was the same in both vessels, but, as the speeds were increased, the turbine vessel gained relatively in economy, and at 18 knots was 15 per cent. more economical, and at $20\frac{1}{2}$ knots 31 per cent., and at full speed 36 per cent. Her superior economy in coal enabled her to reach a speed of 23·63 knots, or $1\frac{1}{2}$ knots more than the *Topaz*, on the same coal allowance. The results of the trials also showed that, at a speed of 20 knots, the *Amethyst* could steam about 50 per cent. more miles than the *Topaz* on the same quantity of coal.

The experience as regards Atlantic liners is as yet limited to three vessels—the *Virginian*, the *Victorian*, and the *Carmania*. The first two are of the Allan line, 520 feet in length, 15,000 tons displacement, and 12,000 horse-power, with a sea speed of from 16 to 17 knots.

These vessels have been running since the spring of 1905, and the consumption of coal has been estimated to be no more, and probably less, than would have been the case had they been fitted with the most economical engines of ordinary type.

The Cunard liner *Carmania*, 672 feet in length, 30,000 tons displacement, and 21,000 horse-power, is a sister vessel to the *Caronia*, propelled by quadruple-expansion engines of the most economical type, and during the last four months the consumption of coal in the two vessels has been carefully measured, but it is too soon as yet to give the results. However, on the official trials, the turbine vessel exceeded the speed of her sister ship by one knot.

Some of the advantages found to exist with turbine propulsion are, that the propellers never race in the heaviest seas, and that, as a consequence, the speed is better maintained under all weather conditions. The cause of this is to be traced to the smaller diameter of the propellers, wider blades, and deeper immersion. There is also much less vibration.

The tendency of late has been to increase the reversing, or astern, power of turbine vessels to such an extent that, in many cases, the stopping and manœuvring powers have been equal to those of twin screw vessels with reciprocating engines. The starting of turbine vessels is relatively quick, for the torsional force of a turbine, when starting from rest with full steam on, is at least 50 per cent. greater than the torque at the usual running speed, because the blades, when running slowly, meet the full blast of the steam instead of moving with it as they do at their usual speeds. With ordinary engines, the starting torque does not exceed the torque at full speed. When manœuvring, turbines cannot fail to respond when steam is turned on, for they have no dead centres upon which to stick, as in the reciprocating engine.

From the fact that the faster and larger the vessel the better has been the performance, it seems safe to infer that the two very large and fast Cunarders now building will give satisfactory results, and the same may be expected as regards new turbine construction in ships of war.

The total horse-power in steamships sailing under all flags is at present about eight millions. Of this total, about one-quarter, or two millions, is in the faster class of ships to which turbines are suitable.

Of the remaining six millions horse-power, about three to four are in the larger class of ocean tramp, and the remainder in coasting steamers and small river boats, etc.

By a combination of the turbine with the reciprocating engine there seems to be no doubt that the three or four millions horse-power of large ocean tramps may be successfully propelled with a saving of from 15 to 20 per cent. in cost of fuel.

This combination has not yet been applied to any vessel. In it the reciprocating engine first expands the steam from the boiler down to about atmospheric pressure, and then passes on to the turbines, which complete the expansion down to the condenser pressure. The turbine thus utilises the lower part of the expansion, which the reciprocating engine cannot do, and the combination is therefore a good one. For manœuvring or stopping the vessel, either the engine or the turbines, or both, may be used, and there seems to be no doubt that this arrangement will come into vogue for the slower class of vessels of larger size.

Turbines have been applied to other uses within the last ten years. The most important of these are for the working of rotary blowers, air compressors, and water pumps.

In general construction the turbine air-blower portion is similar to a steam turbine. The blades or vanes which propel the air are plano-convex in section, and set in rows at an angle similar to that of the blades of a ship's propeller. Between the rows of moving blades are rows of guide blades inwardly projecting from the case. These latter are also of plano-convex section, and are set with their plane surfaces parallel to the axis; and their purpose is to assist the flow, and to stop the rotation of the air after being acted on by the moving blades. Each row of moving and fixed blades adds a little to the pressure, and compresses the air gradually along the annular space between the drum and the case. Balance pistons or dummies are provided for balancing the end thrust of the air, as in the steam turbine. The speed of rotation is 3600 revolutions per minute, and the tip velocity of the air blades about 400 feet per second.

THE EXPANSIVE WORKING OF STEAM
IN STEAM TURBINES

The Watt Anniversary Lecture

Read before the Greenock Philosophical Society,
January 16th, 1909

I find it difficult to add anything to the words of the many illustrious men who have addressed this Society on previous anniversaries of the birth of James Watt—to the words of Sir Humphry Davy, Lord Aberdeen, and Lord Jeffrey, and in later years to those of Joule, Scott-Russell, Preece, and Kelvin. This evening I should prefer to recall to your memories the fundamental principles of steam discovered by James Watt, and to endeavour to trace their application in the engines constructed by him and by the firm of Boulton & Watt, then in the more highly developed forms of compound, triple, and quadruple reciprocating engines, and lastly, in steam turbines on land and sea.

The laws of steam which James Watt discovered are simply these, that the latent heat is nearly constant for different pressures within the ranges used in steam engines, and that, consequently, the greater the steam pressure and the greater the range of expansion, the greater will be the work obtained from a given amount of steam, and secondly, as may seem to us now as obvious, that steam from its expansive force will rush into a vacuum.

Having regard to the state of knowledge at the time, his conclusions appear to have been the result of close and patient reasoning by a mind endowed with extraordinary powers of insight into physical questions, and with the faculty of drawing sound practical conclusions from numerous experiments devised to throw light on the subject under investigation. His resource, courage and devotion were extraordinary, and drew to his side a coterie of kindred spirits with whom he discussed freely his theories and his hopes and the results of his experiments.

In commencing his investigations on the steam engine, he soon discovered that there was a tremendous loss in the Newcomen engine which he thought might be remedied; the loss caused by condensation of the steam on the cold metal walls of the cylinder. He first commenced by lining the walls with wood, a material of low thermal conductivity. Though this improved matters, he was not satisfied: his intuition doubtless told him that there should be some better solution of the problem, and doubtless he made many experiments before he realised the true solution in a condenser separate from the cylinder of the engine. It is easy after discovery to say, "How obvious and how simple", but many of us here know how difficult is any step of advance when shrouded by unknown surroundings, and I can well appreciate the courage and the amount of investigation necessary before James Watt thought himself justified in trying the separate condenser.

But to us now, and to the youngest student who knows the laws of steam as formulated by Carnot, Joule and Kelvin, the separate condenser is the obvious means of constructing an economical condensing engine.

Watt's experiments led him to a clear view of the great importance of securing as much expansion as possible in his engines. The materials and appliances for boiler construction were at that time so undeveloped that steam pressures were practically limited to a few pounds above atmospheric pressure. The cylinders and pistons of his engines were not constructed with the facility and accuracy to which we are now accustomed, and chiefly for these reasons expansion ratios of from two- to three-fold were the usual practice. Watt had given to the world an engine which consumed from 5 to 7 lb. of coal per horse-power hour, or one-quarter of the fuel previously used by any engine. With this consumption of fuel its field under the conditions prevailing at the time was practically unlimited. What need was there, therefore, for commercial reasons, to endeavour still further to improve the engine at the risk of encountering fresh difficulties and greater commercial embarrassments? The course was rather for him and his partners to devote all their energy to extending the adoption of the engine as it stood, and this they did, and to the Watt engine consuming from 5 to 7 lb. of coal per horse-power mankind owes the greatest permanent advances in material welfare recorded in history.

The Watt engine with secondary modifications was the prime mover in most general use for eighty years, till the middle of last century, when the compound engine began to be introduced. Why, we may enquire, was it that the compound engine was so long in coming into use, for it had been patented by Hornblower in 1781, or seventy years before; and why does John Bourne in his large book *Practical Instructions for the Manufacture and Management of every species of Engine*, published 1872,* make no mention in the index of the compound or triple-expansion engine, and when he speaks of Hornblower's double-cylinder engine (really a compound engine), why does he do so in disparaging terms, mentioning that there was no increased economy in steam over the single cylinder? This last statement provides an

---

* "A Treatise on the Steam Engine in its various applications to Mines, Mills, Steam Navigation, Railways and Agriculture. With Theoretical Investigations respecting the motive power of heat and the proper proportions of steam engines. Elaborate Tables of the right dimensions of every part, and practical instructions for the manufacture and management of every species of engines in actual use."

answer to our enquiry, for it is correct in view of the very low steam pressure in general use before that time, or until somewhat before the middle of the last century, when the introduction of the locomotive led to a general rise in pressures on land, and the surface condenser some years later to increased steam pressure at sea. Also we must remember that many experiments have shown that unless the mean difference of pressure on a piston exceeds about 7 lb. per square inch, the friction, the bulk, the momentum of the moving parts, and the cost, make such a cylinder not worth having. The case, however, with the turbine is entirely different, and it is chiefly owing to this difference and to its power of usefully expanding the steam down to the very lowest vapour pressure attainable in the best condenser, that it has surpassed the best reciprocating engines in economy of steam. To return to our subject. The introduction of the compound, triple-, and quadruple-expansion engines was therefore concurrent with the improvements in boiler construction, the introduction of the surface condenser, and the general rise in steam pressure, and by the quadruple engine the expansion ratio has been extended up to about sixteenfold, and the consumption of coal per horse-power reduced to $1\frac{1}{4}$ to $1\frac{1}{2}$ lb. per horse-power hour, or to from one quarter to one third the fuel consumed in the time of James Watt.

Let us now direct our attention to the turbine engine, which derives its power not from the pressure of the steam on a piston but from the momentum of the steam at high velocity curving around and blowing forward the vanes or paddles attached to the shaft. It is unnecessary here to recapitulate the many attempts to construct a successful steam turbine from the days of Hero till a quarter of a century ago, as several excellent books are now published on the subject, of which I may name that by R. M. Neilson. It is true that the difficulties of construction and inferior workmanship available during this early period were a serious bar to progress, but the chief obstacle lay in the fact that the turbine to be economical in steam must (at least in its primitive form) rotate at a very high speed, and that before 1880 there was no commercial use for such a high-speed engine excepting through the intermediary of belts or friction gearing, or for such exceptional purposes as the direct driving of circular saws. The chief purpose for which the turbine is now extensively used on land did not then exist, namely, for the driving of dynamos. Then again, belts for high speeds are a very unsatisfactory appliance, and accurately cut spiral gearing as recently introduced by Dr De Laval had not been devised, and again the problem of applying a turbine to the propulsion of vessels, being surrounded as it was with great consequential difficulties, would naturally only be attacked after the successful application of the prime mover to some easier and simpler purpose on land, so that I think, on the whole, we may safely say that under the conditions prevailing the commercial introduction of the turbine before 1880 was a practical impossibility.

It is a matter of history that the turbine principle had been used for obtaining power from waterfalls before the days of James Watt, but I am not clear that he had in mind any concrete form of steam turbine, yet in 1770 he suggested "a circular engine consisting of a right handed and left handed bottled screw spiral involved in one another", and he also appears to have had a leaning towards some form of directly rotary engine, for in 1769 he took a patent for a Barker's reaction water wheel, the water pressure being derived from the action of steam on water as in Savery's fire engine or a modern Pulsometer. He also designed a rotary abutment engine in 1782, but in none of these machines is there any indication of an attempt to gain greater expansion ratios for the steam.

It is peculiarly interesting to recall on this occasion that one of the earliest steam turbines to be put to practical work was erected in this town about the middle of last century, and was a turbine like that described by Branca in 1629. It consisted of a steam jet playing upon a paddle wheel, coupled to a circular saw, which it drove for some years. The principle of the expansive working of steam was, however, only to a small extent utilised in this engine, for I believe that the steam jet was non-divergent, which implies a useful expansion ratio of only about $1\frac{1}{2}$-fold. One of the most conspicuous workers in the design of the compound turbine was the Rev. Robert Wilson of Greenock, Master of Arts of Edinburgh, reputed to be a man of learning and skilled in many things, who lodged a patent in 1848. This patent is of unusual length and rich in detail, and describes radial flow and parallel flow compound turbines designed for moderate ratios of expansion. The blades and guides were proposed to be fastened by casting them into the hub and case, a method occasionally used at the present time.

The principles of Wilson's designs are generally correct, but the proportions of his turbines are extravagantly incorrect, the blades being too large and too few for success. I had a model made of Wilson's turbine eighteen years ago, and under steam all that could be said was that it went round the right way. I do not think that Wilson can have made a model and tested it before he applied for his patent, the course followed by James Watt, and one which is to be strongly recommended to the attention of inventors generally under almost all circumstances, as saving time, money, and disappointment. There were many workers on steam turbines of English nationality before and since the time of Wilson, but within the last twenty years other countries have taken up the subject with zest.

Prior to 1880 the uses for a very high speed motor were few, as we have seen; the speed of revolution of steam wheels, i.e. velocity of rotation, as

Bourne described them in 1872, "was inconveniently high for most purposes", but after 1880, conditions were changed; the beautiful machine, the milk separator, of Dr De Laval of Stockholm, and the great invention of the dynamo electric machine had come and required a high speed prime mover to drive them, and these provided encouragement to the workers on steam turbines; thus between 1884 and 1888 we find the practical and successful realisation in altered and correct proportions of ideas and suggestions of previous workers, the compound steam turbine in 1884 applied to the direct driving of dynamos, and the single stage impulse wheel in 1888 of very high velocity played upon by the expanding steam jet, both types possessing great ratios of expansion.

All steam turbines now in practical use expand the steam usefully over nearly the whole range from the boiler pressure to the pressure in the condenser, and their designs are based on the principles involved in the construction of their prototypes of 1884 and 1888.

There is first the compound turbine, whose characteristic feature is the gradual expansion of the steam by small drops of pressure at each of a long series of turbines of gradually increasing volumetric capacity, as in the Parsons', or a somewhat less gradual expansion with greater drops of pressure at each stage, as in the Rateau, Zoelly and others.

Then there is the expansion by the divergent jet in one stage, as in the De Laval, or an expansion in a relatively small number of stages by expansion jets playing upon rows of buckets with intermediate rows of reaction guides as in the Curtis and Riedler-Stumpf.

Then there are combinations of the first and second, where the first stage of the expansion is effected by say a Curtis element and the rest by a Parsons, and other combinations have also been proposed, too numerous to mention here.

Let us consider these principal examples of the turbine principle more closely. In the compound turbine the velocities of the steam are low; at each passage through the blades it expands a little, yet it obeys as regards the velocity of efflux approximately the laws of flow of fluids; but the aggregate of the small expansions soon becomes apparent and has to be taken into account when reckoned over a considerable number of the series of elemental turbines; for instance, if the expansion ratio for a single turbine of the series be as 1 to 1·03 in volume, a 3 per cent. expansion, then after passing through 23 turbines the volume of the steam will be doubled and the velocity of flow through the guide blades and moving blades (presuming they are of equal area of passage way) will be about 230 feet per second. The velocity of the blades is, generally speaking, about half the velocity of the steam at issue, and will therefore in this case (which I have taken as common in marine practice) be about 115 feet per second.

The difference in velocity of the steam and the blades is smoothed over largely by the curvature of the blade, which somewhat resembles a shallow hook around which the stream lines in the steam arrange themselves with very little shock or eddying in the steam, so that the coefficient of efficiency is high.

In turbines for driving dynamos, and for other purposes where higher speeds of revolution are permissible, steam velocities up to 600 feet and blade velocities up to 300 feet per second at the exhaust ends are general.

In turbines of the Rateau, Zoelly, and other types with multiple discs, each disc carrying one row of blades only and working in a cell through the walls of which the shaft passes in a steam packed gland, nearly the whole drop in pressure takes place at the guide vanes, and very little at the moving vanes, which are of cup form. Here the velocities of the steam generally range from 900 to 1100 feet per second and the velocities of the blades from 350 to 450 feet per second. In turbines, however, of the De Laval single-wheel, and of the Curtis and other types with a relatively small number of pressure stages, higher steam velocities are used, ranging from 4200 feet per second in the single-wheel down to 1500 feet in a seven-stage Curtis turbine. The jets used in the single-stage turbine are of very divergent form, but when the expansion is divided over seven stages, very little divergence is necessary. In the single-stage turbine blade velocities as high as 1200 feet per second are adopted, the discs being of taper form and of the strongest nickel steel. But even this high velocity is insufficient to obtain a very good coefficient of efficiency from the steam, and when the disc is made large so as to reduce the immense angular velocity incidental to the high peripheral speed, the skin friction of the disc and the prime cost and weight increase rapidly.

In the Curtis five-stage, the blade velocities are about 460 feet per second and the steam velocity about 2000 feet per second, and by its passage through two rows of moving and one row of guide blades the steam is brought nearly to rest before passing on to the next succeeding chamber. By this sinuous treatment of the steam efficiencies are obtained comparable to those of the compound turbine.

From the commencement of turbine design in 1884 I have avoided the adoption of high steam velocities on account of their cutting action on metals when any water is present. The cutting has been found to be due not to the impact of gaseous steam, but to that of minute drops of water entrained by the steam, and hurled by it against the surfaces. The drops, formed like fog consequent on the expansion of saturated steam, are sufficiently large to cause the erosion. To test the effect in an extreme case, a hard file was placed opposite to a jet of steam issuing at 100 lb. pressure into a vacuum of 1 lb. absolute pressure. In 145 hours it was found to be eroded to the extent of about 1/32 inch as if it had been sand blasted. The calculated velocity of the issuing steam in this case was about 3800 feet per second, and the striking fluid-pressure of a drop of pure water at this velocity about 90 tons per square inch. Owing, however, to the receding velocity of the blades from the blast in all turbines the erosive effect is much reduced. In multicellular turbines of few stages, though the erosion is slow, yet provision is necessary for renewal of blades at intervals. In turbines of many stages it is still slower, and in the compound turbine erosion is, practically speaking, absent, and renewal of blades unnecessary. This absence of the tendency to erosion in compound turbines permits the use of brass or copper blades, which are found to preserve their polish, are not liable to corrosion or rusting, and preserve their smoothness of surface and the initial economy of the engine unimpaired for many years.

It is now just fifteen years ago, and exactly ten years from the commencement of work on the compound steam turbine, that the results obtained on land were thought to justify an attempt to apply the turbine principle to the propulsion of vessels. These results lay in the fact that a condensing turbine engine of 200 horse-power with an expansion ratio of 90 volumes had been found to have equal economy to a good compound piston engine, and that, moreover, there were within sight reasons to hope for still better results. A commencement was made, and by the end of 1897, after three years of hard work and experiment, the *Turbinia* was completed. Her trials were usually made on the measured mile in the North Sea, but occasionally, when the sea was too rough, runs at speeds up to 31 knots were made on the Tyne, where the legal limit of speed of steamships was 7 knots. Thus by the magnanimity of the Tyne Improvement Commissioners the completion of the *Turbinia* was greatly facilitated, though it is fair to say that great care was exercised and no harm was done to the public. In her the problem of adapting the turbine to the screw propeller was worked out. The result was a compromise between the two. The turbine had to be made short and broad so as to revolve as slowly as possible, and the screw had to be made with finer pitch and wider blades. The result in propulsive efficiency was found to be good, and the problem satisfactorily solved for fast vessels of 16 knots and upwards. It was also seen that the faster the vessel the more favourable would be the economy of the turbine as compared with the reciprocating engine.

The destroyers *Viper* and *Cobra* followed. The next step was the application of the turbine to vessels of commerce.

Dumbarton was the scene of many conferences. Mr Archibald Denny was deeply interested in the problem, and so was Captain John Williamson, with the result that the first passenger vessel, the *King Edward*, was built in 1901 at Dumbarton to the joint ownership of Captain John Williamson, Messrs Denny, and the Parsons Marine Steam Turbine Co., Ltd. The success of this vessel soon led to the adoption of turbines in cross-Channel steamers, and also led, aided by the success of the destroyers *Viper* and *Velox*, to the specification of turbines in H.M.S. third-class cruiser *Amethyst*. From that time turbines began to be rapidly adopted for fast vessels, including the largest and fastest mercantile and war vessels afloat.

The success of the *King Edward* in 1901 was a red-letter day for the marine turbine. Let us enquire in what this success consisted. In the first place, a factor of primary importance is the coal bill, and it was soon proved by Messrs Denny that this was less to the extent of from 15 to 25 per cent. than with vessels propelled by reciprocating engines of equal dimensions and boiler capacity. Also the cost of oil, which with reciprocating engines amounts to about 5 per cent. of the coal bill, was nearly eliminated. The vibration was also less. Then the upkeep of machinery was found to be favourable, and as the crew became accustomed to her the coal consumption still further diminished, and I am informed by Captain Williamson that this further decrease has been well maintained up to the present time. The exceptional reliability of the machinery also became more and more assured.

There are now about 120 vessels actually on service fitted with turbines, and 70 more under construction, representing a total horse-power of marine turbine engines of about 2,250,000 of which 1,250,000 horse-power is completed.[1]

Two other great steps in the adoption of the turbine occurred almost simultaneously in 1905, namely, the decision of the Admiralty to adopt turbines for all new construction in fighting ships and the adoption of turbine machinery for the great Cunarders. The steps from the third-class cruiser *Amethyst* of 14,000 horse-power to the *Dreadnought* of 22,000, and to the *Indomitable* of 41,000 were, it is true, more gradual, but the number of vessels involved was great. In the Mercantile Marine the step from the *Queen*, the first cross-Channel vessel of 8000 horse-power, directly to the *Lusitania* and *Mauretania* of 70,000 horse-power required great courage on the part of the late Lord Inverclyde and his co-directors and engineers. Such steps as these are not taken without thorough investigation based on ascertained results. When it is considered that the low-pressure turbine in the *Queen* was 6 feet in diameter, 20 feet in length, and 25 tons in weight, as compared with the Cunarders' low-pressure turbines of about 17 ft. 6 in. diameter, 50 feet in length, and 300 tons in weight, it is realised what a great departure was involved. Forces and conditions were altered, and differential expansions and deflections of the structure had all to be reconsidered in detail, for though they had been successfully

---

[1] By May, 1909, these figures had risen to 273 vessels and a total horse-power of 3,530,000, built and under construction.

dealt with and controlled in the smaller engine the magnitude of the larger structure rendered re-calculations and thorough investigation necessary; thus no room was left for the possibility of any adverse conditions arising owing to the very great increase in the size of structure, and everything that care, thought and experience could accomplish was done. The results have satisfactorily agreed with the hopes and estimates of all concerned.

But to return to the subject of our lecture. In the *King Edward* there was an increase in the ratio of expansion beyond that hitherto realised in any reciprocating engine. Her boiler pressure is 150 lb. The pressure at the inlet to the turbines at normal full speed is 130 lb., the pressure in the condenser is $1\frac{1}{2}$ lb. absolute, giving a ratio of expansion of 87 by pressure or about 66 by volume, as compared with the volumetric ratio of about 10 in triple-expansion reciprocating engines for a similar class of vessel.

In some later turbine vessels higher steam pressures have been adopted, resulting in a small gain in efficiency, partly counterbalanced by the greater weight of the turbine cases, and, if the vessel has Scotch boilers, then also by the greater weight of the boilers to carry the greater pressure. On the whole the net gain, if any, is but small.

A substantial increase in efficiency has however been realised by improvements in condensers and pumps, in order to take full advantage of the property of the turbine of expanding steam usefully to the lowest pressure attainable in the condenser. Before the turbine came into use a very high vacuum was not found desirable, for the simple reason that the reciprocating engine is unable to utilise it. For instance, a triple-expansion engine does not gain in economy of coal if the absolute pressure in the condenser be diminished below $2\frac{1}{2}$ lb. The turbine, however, derives a net gain in efficiency of 13 per cent. from a diminution of pressure in the condenser from $2\frac{1}{2}$ lb. absolute to 1 lb. absolute. The improvements that have been introduced of late years in condensing plants consist primarily in improved design of the condenser, and in improvements in air pumps to increase their volumetric capacity. In the condenser the tubes are so spaced and grouped that the steam, attenuated into relatively an enormous volume, shall pass freely without much resistance and drop of pressure throughout the whole surface, and provision is made by the form of the condenser shell, with or without a single baffle plate, so that the suction of the air pump shall remove the air uniformly from all parts. The vacuum now usually obtained in well-equipped turbine vessels is very close to that corresponding to the temperature of the circulating water leaving the condenser. The difference is sometimes as small as two degrees, so that there is no room for much further improvement in this direction. To increase the volumetric capacity of the air pumps, dry air pumps run at a high speed may be used, separate pumps being employed to remove the water of condensation. An alternative, and perhaps a preferable method, is the Vacuum Augmentor, a simple apparatus without moving machinery, which consists of a very small steam jet placed in a narrowed portion of the ordinary air-pump suction, which sucks the air out of the condenser and compresses it through a small intermediate cooler into the suction of the air pump, the water of condensation draining by gravity through a water seal into the same air-pump suction.

Further possible improvements would therefore seem to tend in the direction of an increase in the efficiency of the turbine itself. In large turbine vessels the ratio of the shaft horse-power to the total available energy in the steam from boiler to condenser reaches 70 per cent., and the question is whether there is a probability of somewhat reducing this loss of 30 per cent. During the last eleven years a small reduction in steam per horse-power delivered to the shaft has been brought about by minor improvements in design, better finish and proportion of the blading, and by the increased size of the engines constructed.

In 1897 the *Turbinia* consumed 16 lb. of steam per shaft horse-power per hour for all purposes; in 1901 the *King Edward* consumed 16 lb. per shaft horse-power for all purposes; in 1907 the *Lusitania* consumed 14·94 lb. per shaft horse-power for all purposes.

In the case of slow vessels where the exigencies of the screw propeller limit the revolutions to a low rate I have for many years advocated a combination or partnership between the reciprocating engine and the turbine, which seems to promise a high degree of efficiency and to suit all the requirements of the case. In this combination each engine deals with that part of the expansion for which it is best suited, the reciprocating engine taking the high pressure portion from the boiler pressure down to about atmospheric pressure, and the turbine carrying on the expansion from about atmospheric pressure right down to the condenser pressure.

The reciprocating engine is thus relieved of the low pressure part of the expansion, which at best it carries out in a very inefficient manner, losing as it does all the last part, and the turbine is relieved from the high pressure part, which when constructed for slow revolutions it performs unsatisfactorily. But the turbine designed for low pressures and slow revolutions is an engine which converts a very high percentage of the power in the steam into the shaft horse-power.

Messrs Denny have fitted the *Otaki* of 8000 tons, 500 horse-power and 13 knots sea speed with this system, the boiler pressure being 200 lb., no superheaters being fitted, and the very low consumption of 12·3 lb. of steam for all purposes was registered on trial. Messrs Harland & Wolff are also fitting a vessel for the Dominion Line on this system.

James Watt, we are told, suggested the screw propeller in 1770, half a century later it commenced to come into use, and now it is almost universally adopted in all new construction.

It is a very interesting and curious fact to note that in the first instance, and for many years, the screw was driven by spur gearing from a very slow speed engine, presumably because the builders of engines were afraid to design the engines to run as fast as the screw required to be driven. Now for forty years or more gearing has been entirely abandoned, and the high speed reciprocating engine has worked well.

The turbine has now come on the scene and its best speed of revolutions is faster than that of the screw, excepting in fast vessels. For the larger portion of the tonnage of the world it is at present unsuited, except to take a secondary but excellent part in the combination system.

We may naturally speculate as to the future, and enquire if there is a possibility of the turbine being constructed to run more slowly, and without loss of economy, or whether the propeller can be modified to allow of higher speed of revolution.

Or, again, may a solution be found in reverting to some description of gearing, not the primitive wooden spur gearing of half a century ago, but steel gearing cut by modern machinery with extreme accuracy and running in an oil bath; helical tooth gearing, or chain gearing, or, again, some form of electrical or hydraulic gearing?

These are questions which are receiving attention in some quarters at the present time, and if a satisfactory solution can be found, then the field of the turbine at sea will be further extended.

THE APPLICATION OF THE MARINE STEAM TURBINE AND MECHANICAL GEARING TO MERCHANT SHIPS

Institution of Naval Architects, March 18th, 1910

The steam turbine has not, as yet, been applied to vessels of slow normal speed on account of the high initial cost and inferior economy in steam; further, no promising scheme has, as yet, been evolved having for its object the modification of the turbine or propeller, so as to reduce the efficient speed of revolution of the former, and increase that of the latter for vessels of 12 knots sea speed and under, and the only approach to meeting these conditions (if we except gearing propositions) has been in the combination system, where the turbine plays a secondary part in the equipment, by utilising the lower portion of the expansion of the steam between the low-pressure cylinder of the reciprocating engine and the condenser.

The complete and most satisfactory solution for slow-speed vessels would appear to be by means of gearing, provided the losses in transmission, first cost, and cost of maintenance are not too great. Many forms of gearing—mechanical, electrical, and hydraulic—have been proposed or applied on a small scale.

I believe the first application of helical spur gearing to drive a propeller was made by the Parsons Marine Steam Turbine Company, Limited, in 1897. The turbine was of 10 horse-power geared to two wheels, each wheel driving a propeller shaft. The revolutions of the propellers were 1400 per minute, and the ratio of the gear 14 to 1. The turbine was of the Parsons type, with a reversing turbine on the same shaft incorporated in the same casing. The gear was single helical. The turbine took part of the thrust of the propeller, the remaining thrust being taken on the thrust bearing in the gear casing. The air, circulating and oil pumps were driven by worm gearing off one of the screw shafts. The launch was 22 feet over all, and attained a speed of 9 miles an hour. She was built to the order of Mr F. B. Atkinson, for his yacht *Charmian*. The launch was sent to the Turbinia Works in 1904, and the turbine was generally overhauled and cleaned. The gear was found to be in perfect order, and did not require any repair.

Helical and double helical gear of fine pitch suited to high speeds of rotation was, I believe, first introduced by Dr De Laval, of Stockholm, and has been extensively used in connection with his turbine for many years with entire success, and at a moderate cost of maintenance. I have had several experimental sets constructed. One of these was a double helical gear of the De Laval type made in 1897, gearing from 9600 revolutions of the turbine to 4800 of the dynamo, transmitting 300 horse-power. The efficiency was estimated by the method of heat loss to be above 98 per cent. This gear was cut in an ordinary universal milling machine without any special precaution as to accuracy, and I was much impressed (in spite of the obvious irregularity of the teeth) by finding how well it ran, except that it made a considerable noise.

In the summer of last year the directors of the Turbinia Works Company decided to test turbines mechanically geared to the screw shaft of an existing typical slow-speed vessel, and a cargo vessel named the *Vespasian* was purchased for this purpose.

The *Vespasian* was built in 1887 by Messrs Short Brothers, of Sunderland. Her dimensions are: length on load water line, 275 feet; breadth moulded, 38 ft. 9 in.; depth moulded, 21 ft. 2 in.; mean loaded draught, 19 ft. 8 in. and displacement, 4350 tons.

Previous to installing turbine engines with reduction gear, the vessel was fitted with an ordinary triple-expansion surface-condensing engine by Mr G. Clark, of Sunderland, with cylinders 22¼ inches by 35 inches by 59 inches and 42 inches stroke. The air, circulating, feed and bilge pumps were driven from the intermediate-pressure crosshead, with the usual arrangement of levers and links. The condenser was cast with the back columns of the main engine, and had a cooling surface of 1770 square feet. The boilers —two in number—are 13 feet diameter by 10 ft. 6 in. long, with a total heating surface of 3430 square feet, and grate area of 98 square feet, working under a pressure of 150 lb. with natural draught. The propeller is of cast iron, and has four blades, having a diameter of 14 feet, pitch 16·35 feet, and expanded area of 70 square feet.

With a view to obtaining comparative data between the turbine installation and the reciprocating engine, it was decided to run trials of the vessel with her reciprocating engine, prior to its removal and the installing of the turbines and gearing.

Before proceeding on the voyage from which data regarding the performance of the reciprocating engine were taken, the propelling machinery was completely dismantled and overhauled. The high-pressure piston valve chamber was rebored and new valve rings fitted; slide valves were replaned and faced up; bearings were renewed, and other repairs carried out wherever necessary to bring the machinery into an efficient condition and first-class working order. To obtain reliable measurements of water consumption, two tanks were fitted, each of 400 gallons capacity, with suitable change cocks and connections for the air pump to discharge through these measuring tanks.

It was necessary, for the purpose of obtaining data under service conditions, that the vessel should be run in her loaded condition. Arrangements

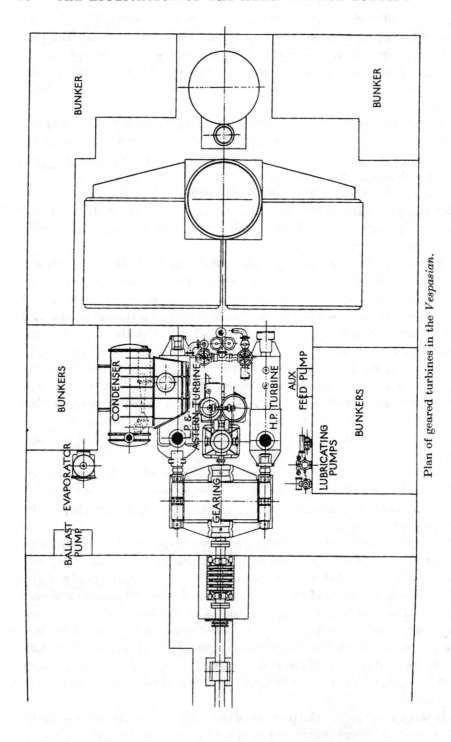

Plan of geared turbines in the *Vespasian*.

were consequently made with a local firm of shipbrokers to take a cargo of coal from the Tyne to Malta, and on June 26th, last year, the *Vespasian* left the Tyne in a loaded condition with a special recording staff on board, and on this voyage careful measurements of coal and water consumption were made.*

On the completion of the voyage, the vessel returned to the Turbinia Works, where her reciprocating engine was taken out, engine seats re-modelled, and preparations made for the reception of the turbines and gearing. [A plan of the turbine machinery, taken from a plate in the paper, is reproduced herewith.]

The only alteration made to the vessel was in the type of propelling engines; the boilers, propeller, shafting and thrust blocks remained the same as for the reciprocating engine.

The propelling machinery consists of two turbines in "series", viz. one high-pressure and one low-pressure, the high-pressure turbine being placed on the starboard side of the vessel and the low-pressure on the port side. At the after end of each of the turbines a driving pinion is connected, with a flexible coupling between the pinion shaft and the turbine, the pinion on each side of the vessel being geared into a wheel, which is coupled to the propeller shaft.

A reversing turbine is incorporated in the exhaust casing of the low-pressure turbine. The air, circulating, feed and bilge pumps are of the usual design for tramp steamers, and are driven by means of a crank and connecting rod coupled to the forward end of the gear-wheel shaft. The turbine and pinion shaft bearings are under forced lubrication, similar to ordinary turbine practice. The teeth of the pinions and of the gear wheel are lubricated by means of a "spray" pipe extending the full width of the face of the wheel. Independent oil pumps are fitted for supplying oil to the bearings and gear wheel, with a view to the possibility of experimenting with different lubricants for the gear wheel, the oiling system for the bearings being separate from that of the gear wheel.

The high-pressure turbine is 3 feet maximum diameter by 13 feet overall length, and the low-pressure 3 ft. 10 in. diameter by 12 ft. 6 in. length. The turbines are similar in design to a land turbine, being balanced for steam thrust only, the propeller thrust being taken up by the ordinary thrust-block of the horse-shoe type, which is fitted aft of the gear wheel.

A new condenser, together with a vacuum augmentor, was fitted with the turbine installation. The cooling surface of the condenser is 1165 square feet.

The gear wheel is of cast iron, with two forged steel rims shrunk on. The

* The data and results of a progressive trial carried out on the Whitley Bay mile are given and plotted in an Appendix.

diameter of the wheel is 8 ft. 3½ in. pitch circle, having 398 teeth—double helical—with a circular pitch of 0·7854 inch. The total width of face of wheel is 24 inches; inclination of teeth 20° to the axis.

The pinion shafts are of chrome nickel steel, 5 in. diameter pitch circle, with 20 teeth 0·7854 circular pitch. The ratio of gear is 19·9 to 1.

On the completion of the erection, on board, of the turbine-gearing installation at the end of February of this year, the vessel was loaded to the same draught and displacement as on the preceding trials. As already mentioned, the propeller has not been touched or altered in any way. In the short interval since the completion of her alterations, the vessel has been out to sea on four occasions.

The following table gives the data and results of a run made off the Tyne on the 11th of this month (March, 1910) at varying revolutions:

Speed in knots	8·4	9·56	10·5	10·66
Revolutions per minute	56·5	65·0	71·3	73·3
Boiler pressure	145	144	140	145
Initial pressure high-pressure turbine	60 lb.	86 lb.	110 lb.	121 lb.
Initial pressure low-pressure turbine	15·2 in.	12·5 in.	7·1 in.	5·5 in.
Vacuum	28·8 in.	28·8 in.	28·7 in.	28·5 in.
Barometer	29·9 in.	—	—	—
Shaft horse-power	456	740	980	1095
Water consumption per hour, main engines	9070 lb.	12,000 lb.	14,480 lb.	15,670 lb.
Water consumption, all purposes ...	9670 lb.	12,620 lb.	15,120 lb.	16,370 lb.
Water consumption per shaft horse-power, main engines	19·8 lb.	16·2 lb.	14·8 lb.	14·3 lb.

The water consumptions per hour at the several rates of revolutions have been plotted on the diagram, shown in full lines for the reciprocating engines and in dotted lines for the turbine-geared engines.

Under normal full speed steaming conditions an increase of about one knot is obtained with the same coal consumption.

The observed mean speeds on the measured mile given in the above table correspond to the speeds obtained with the reciprocating engines at the same revolutions, thus eliminating any necessity for allowances, the weather conditions in the two cases being very similar.

It may be mentioned that the turbines and gearing have given no trouble, and have worked satisfactorily with very little noise or vibration throughout the trials. Further, there is no appreciable wear on the teeth or bearings.

It is proposed to put the vessel into commission and run extended trials.

I should like to add that the weight of the reciprocating installation was about 100 tons, and the turbine installation 75 tons, saving about 25 tons in engine-room weights. Hence at normal speed of revolution, viz. 65 revolutions in the old engines, and also at about 70 revolutions which would be maintained as the normal speed with the gearing, assuming the efficiency of the gear is 98½ per cent., as determined by experiments, the ratio of shaft horse-power to indicated horse-power with the reciprocating engines works out at about 90 to 91 per cent. No data are given of shaft horse-power taken with the reciprocating engines because the observations

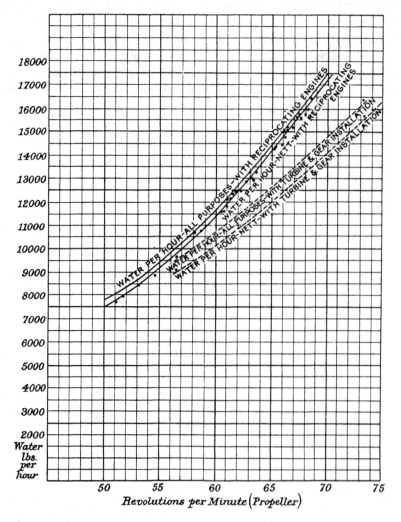

were found to be unreliable, but with the turbine installation accurate shaft horse-powers were obtained. These data of shaft horse-power obtained under identically the same conditions of vessel, propeller, revolutions and speed on measured mile, with both kinds of machinery installation, supply the shaft powers more accurately than could have been obtained under any circumstances from dynamometer readings with the reciprocating engines, where the variation of the torque over the revolution was great, varying at some speeds nearly as 3 to 1. With the turbine installation, on the other hand, the torque is absolutely constant over the revolution, and for this and other reasons I would venture the prediction that with geared turbine installations broken screw shafts will be unknown.

Replying to a written enquiry by Mr C. I. Davidson, Mr Parsons said "The probable behaviour of geared turbines in a heavy sea has been carefully considered. The chief features of the case are that the momentum of the turbines is so enormous, compared with that of the reciprocating engines and propeller, that even without a governor of any kind the acceleration above normal speed of revolution during the period of the screw being out of water is trivial when compared with that of reciprocating machinery. If governors are fitted, these will have ample time to act, and the acceleration will then under no circumstances exceed double (or about 10 per cent.) that of land turbines driving dynamos when the whole load is thrown off instantaneously. This is comparable with from 100 to 300 per cent. acceleration of reciprocating machinery in a heavy seaway."

EXPERIMENTS ON THE COMPRESSION OF LIQUIDS AT HIGH PRESSURES

(With S. S. Cook)

Read before the Royal Society, May 25th, 1911

Introduction

During the experiments on the behaviour of carbon under high pressures and temperatures, of which an account was given in the paper read before this Society by the Hon. C. A. Parsons on June 27th, 1907, very considerable volumetric compressions were observed, and the apparatus then employed appeared to be equally suitable for the direct measurement of the compressibility of liquids at higher pressures than had previously been attempted.

The measurement of the compressibility of liquids has received the attention of a great many investigators; the most comprehensive researches in this subject appear to be those of Amagat, who determined the coefficients of compressibility of water and ether for pressures up to 3000 atmospheres and for a variety of temperatures. Amagat's experiments are given in various numbers of *Comptes Rendus*.* In the experiments about to be described, the pressures were carried up to about 40 tons per square inch, or over 6000 atmospheres.

The experiments were commenced in 1908, but, the first experiments showing some modification of the apparatus to be desirable, principally with a view to making temperature measurements, it is the results of the later experiments carried out during the last year that are chiefly given. A brief reference is desirable, however, to the preliminary readings obtained, by way of explaining the method adopted to eliminate friction, and in order to indicate how these experiments suggested an extension of the research to the investigation of the effects of internal molecular forces of the liquids tested.

Apparatus

The apparatus consisted of a steel mould of about 4 inches bore and 12 inches external diameter, placed under a heavy hydraulic press capable of exerting a pressure of 2000 tons, with a main ram of 29 inches diameter and a 6-inch lifting ram. In the first series of these experiments the mould was constructed of gun steel, having an elastic limit of about 40 tons per square inch. Greater strength could have been obtained, and still higher

---

* *Comptes Rendus*, Vol. ciii, p. 429; Vol. cv, p. 1120; Vol. cvii, p. 522; Vol. cviii, p. 228; Vol. cxi, p. 871.

pressures applied, by the employment of special steel with a higher elastic limit.

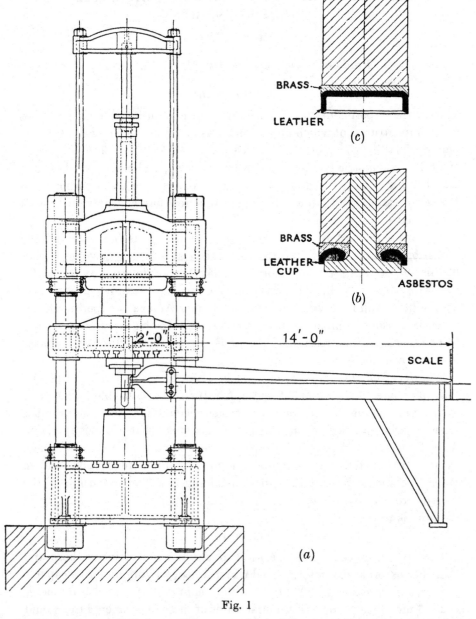

Fig. 1

Water was supplied to the upper side of the ram at pressures up to 2000 lb. per square inch by a three-throw hydraulic pump driven by an electric motor. The pressure on the ram was recorded by a carefully calibrated Bourdon gauge, and the depression of the liquid in the mould measured by a pair of multiplying callipers, inserted between the top of the mould and a projecting collar on the plunger. One of the outer arms of the callipers moved over a graduated scale attached to the other arm, and the readings of this scale were for safety observed through a telescope outside the armoured building in which the press and mould were set up. A sketch of the apparatus is given in Fig. 1 (a). Between the plunger and the liquid was placed a leather cup backed by a thin-edged cup of brass, this combination making an effective packing (see Fig. 1 (b)).

ELIMINATION OF ERRORS

In such an apparatus certain errors would arise in a single experiment, but the experiments were carried out in such a manner as practically to eliminate them. The sources of error were the friction of the packing, the compression of the cup leather and brass and of the plunger below the point of measurement, and the lateral expansion of the mould. All these, with the exception of that due to friction, would be eliminated if an experiment could first be made with the mould filled with an incompressible fluid, and the apparent compressions so obtained deducted from those obtained in any other experiment. But as there was no suitable fluid of negligible compressibility, in order to obtain in an indirect manner the apparent bulk compression in this apparatus of a substance of negligible compressibility, the following procedure was adopted. A steel cylinder was prepared of half the volume of the liquid and of smaller diameter than the bore of the mould. In a first experiment this steel cylinder was inserted in the mould, which was then filled with distilled water up to a mark corresponding to a total volume of 2000 c.c., so that it then contained 1000 c.c. of water and an immersed volume of 1000 c.c. of steel, both of which could be subjected together to bulk compression, and the apparent compression observed. A second experiment was then made with the steel cylinder removed, and the mould filled to the same mark with 2000 c.c. of distilled water, the apparent compression being read as before.

Since the volumetric compression of the water in the first experiment is only one-half of its value in the second (the original volume of the water compressed being only one-half), by subtracting the compression readings of the second experiment from twice those of the first, the compression of the water is eliminated, and the difference thus obtained gives the constant error of the apparatus on the supposition that the inserted steel cylinder is incompressible.

A small correction is then necessary for the compression of this steel cylinder. The coefficient of compressibility of steel at ordinary pressures, as deduced from the modulus of rigidity and Young's modulus, by the usual formula for the relation between the moduli,* is 0.78×10^{-6} per atmosphere of pressure. Since this is only a small percentage of the compressibility determined below for water, etc., the reduction in volume of the steel cylinder under a pressure of p atmospheres may be assumed, to a near enough approximation, to be cp times its original volume, where $c = 0.78 \times 10^{-6}$. In estimating the compressibility of water by the difference between the apparent compressions obtained in the two experiments, it is obvious that it is greater by approximately this amount (viz. 0.78×10^{-6}) than it would be on the assumption of absolute constancy of volume of the immersed steel cylinder.

A further small correction has been necessary owing to the expansion of the mould taking effect upon slightly different volumes in different experiments. This correction has been estimated by the usual theory of the bursting strains of thick cylinders, its magnitude at the highest pressures employed never exceeding 1 per cent.

ELIMINATION OF THE EFFECTS OF FRICTION

Curve 1 gives the readings without corrections obtained with the mould containing 2000 c.c. of distilled water at 18° C. The vertical ordinate represents the volume of the liquid as indicated by the depression of the plunger, and the abscissa the pressure in atmospheres.

The cyclic nature of the curve will be noticed, the upper branch of the loop being plotted from the readings obtained when the pressure was increasing, and the lower from those obtained whilst the pressure was being reduced; the difference between the readings of the upper and lower branches is concluded to be due to friction, the horizontal breadth of the loop being twice the pressure needed to overcome the friction of the packing and of the working parts of the press.

Loops of this nature were traced for all the experiments. To guard against leakage, after the pressure had been raised to its highest value and reduced back to zero, it was again raised so as to repeat a portion of the higher branch of the loop; if any leakage had taken place it was thus at once detected, and the experiment rejected.

By drawing a curve midway between the two branches of the loop we obtain a curve of compression with the effects of friction eliminated. Curve 2 is the curve plotted for 2000 c.c. of water in this manner. Curve 3 is a curve of errors of the apparatus obtained by a calibrating experiment with the immersed steel cylinder as described above.

Curve 4 gives readings plotted for a heavy cylinder oil of mean density about 0.89 at 20° C. It will be noticed that the loop of Curve 4 is very similar to that for water, Curve 1, in spite of the greater viscosity of the

* See Thomson and Tait's *Natural Philosophy*, § 683.

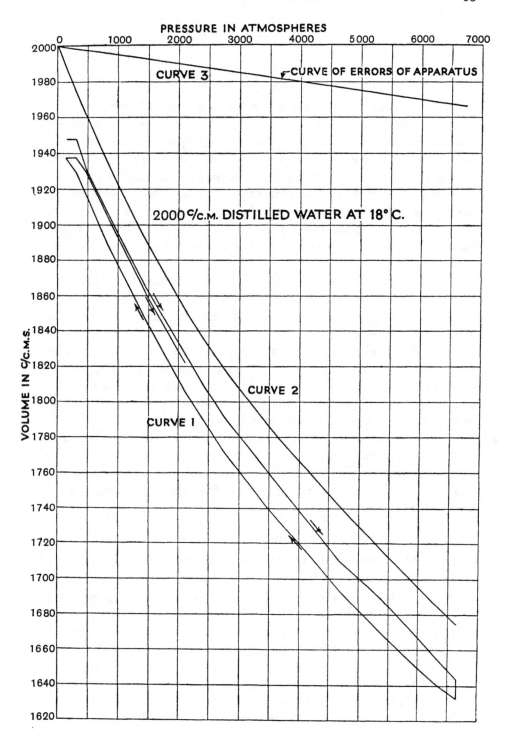

oil, so that it would appear that the loop obtained in all the experiments with liquids is entirely due to friction of the press and packing, and not to internal friction.

BEHAVIOUR OF GRAPHITE UNDER PRESSURE

Similar measurements were attempted with powdered Atchison graphite of specific gravity 2·23 and containing 2¾ per cent. of ash. The readings obtained are plotted in Curve 5. It was thought possible that under high pressure graphite would behave as a fluid, but the character of the curve obtained indicates an increasing degree of consolidation as the pressure is increased. The consolidated graphite appears to show considerable rigidity, that is to say, the stress was not equal in all directions. The lower portion of the cycle was repeated, applying and removing the pressure a great number of times, the dotted Curve 6 representing the readings subsequently obtained.

On opening up the mould the graphite was removed in biscuit-like fragments, often separating along a conical surface at about 45° to the axis, so that it appears to have relieved itself during expansion along lines of cleavage at this angle.

From these results it was difficult to conclude whether the graphite had undergone much bulk compression. The bulk compressibility was, however, subsequently determined by the compression of 1400 grammes immersed in 1300 c.c. water, and was found to be about $6·5 \times 10^{-6}$ in atmospheric units. Curve 7 is a curve of bulk compression deduced in this manner for comparison with Curve 6, and, as will be seen, is nearly parallel to the upper branch of the loop.

Similar results were obtained with Ceylon graphite of specific gravity 2·3 and containing about 15 per cent. of ash. The readings for this graphite are plotted in Curve 8.

EFFECTS OF HEAT OF COMPRESSION

It will be observed that in the curves plotted for the experimental readings, the readings for the pressures ascending a second time for the purpose, as explained above, of detecting leakage, are in general very slightly below the readings for the previous application of the pressure. Subsequent examination of the apparatus showing no trace of leakage, and this phenomenon persisting in subsequent experiments, it was concluded to be due to temperature variation, the work done during compression transforming into heat and rendering the substance, whilst under high pressures, slightly hotter than its surroundings, so that at the end of the cycle it had lost some heat to the surroundings, and during the second compression was slightly colder than during the first. This heat effect being therefore a possible source of error, it was desirable to make some measurement of its amount.

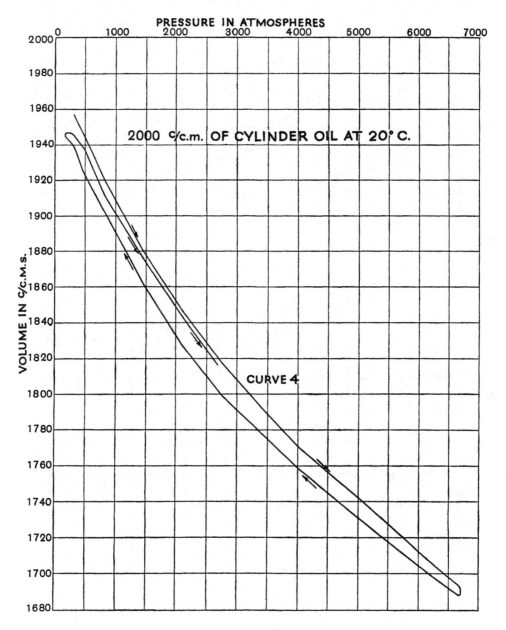

As a preliminary, a volume of 2000 c.c. of heavy cylinder oil was compressed to a pressure of 6300 atmospheres, which was maintained for two hours to allow the heat due to compression to leak away. At the end of this time the pressure was suddenly released, and the liquid was found to be

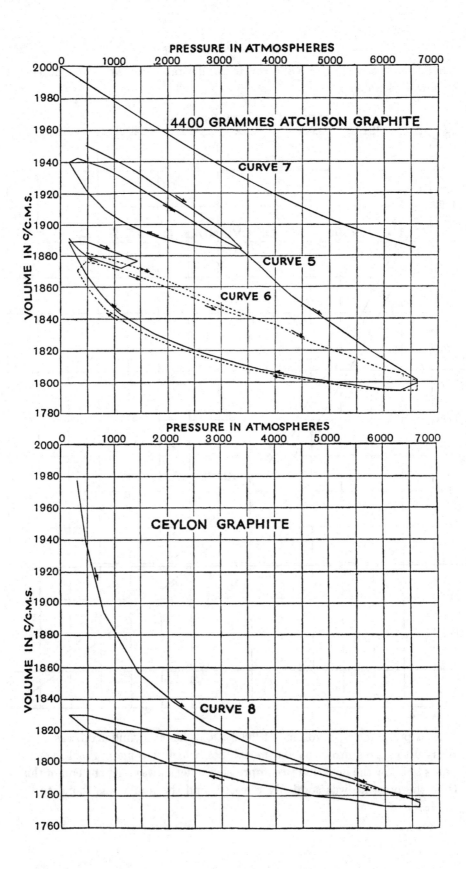

$\frac{1}{8}$ inch lower in the mould than it was before compression, corresponding to a reduction in total volume of about 2 per cent.

The heat effect was further investigated by allowing 2000 c.c. of thin machine oil to cool under high pressure for two hours with an external temperature of about 11·5° C., and after sudden removal of the pressure opening up the mould for examination. With the method of packing employed at this time (see Fig. 1 (b)) it was not possible to open up in less than about a quarter of an hour, owing to the difficulty of removing the brass cup and the cup-leather from the mould.

The temperature after opening up was read, however, at regular intervals of time, by which means, assuming a logarithmic law, the temperature immediately after release and expansion could be estimated, and appeared at the moment of release to have been about 24·5° C.

It was evident from this that the loss of heat was considerably in excess of that due to the work done on the piston during expansion. The specific heat of the oil was measured and found to be about 0·5. The temperature reduction, 36° C., which took place corresponds, therefore, to 18 Centigrade heat units per lb. The work done by expansion was about 42,000 ft. lb. for a mass of about 4 lb., which is equivalent only to 7·5 heat units per lb., leaving 10·5 units per lb. to be accounted for by internal forces.

This point having been reached, the experiments were continued with a twofold object, namely, (1) to ascertain the compressibilities of various liquids for both adiabatic and isothermal compression, the experiments last referred to showing it to be necessary to discriminate between these modes of compression, and (2) to determine the amount of the heat developed from internal forces during compression and absorbed by the same agency during expansion.

Various attempts were made to measure the instantaneous temperature of the contents of the mould during compression by electrical methods with platinum coils and terminals brought outside the mould to a galvanometer. None of these were successful; the readings suffered interference from galvanic action and from conduction in the liquid, and constant difficulty was experienced through breakage of the coils, with congealed oil, and through leakage of the fluid past the terminals, in spite of many precautions taken to avoid these difficulties.

In the meanwhile an accumulator had been installed, and it was now possible to put on or remove the full pressure almost instantaneously; further, by attaching the packing to the ram in such a way that it could be withdrawn immediately on the removal of pressure (see Fig. 1 (c)), it was possible to insert a quick-reading mercury thermometer immediately after release. The mould was also surrounded by a water jacket so that the temperature could be varied as required and measured by a mercury thermometer.

Isothermal curves were plotted by allowing the pressure to remain applied long enough for the liquid in the mould to assume the temperature of the jacket, after which the pressure was varied slightly above and slightly below its recorded value sufficiently to overcome the friction.

For adiabatic compression the pressure was raised to the required value in a few seconds. In these later experiments the pressures were not carried higher than from 4500 to 5000 atmospheres, the adoption of which lower values, together with the experience acquired in working with these high pressures, rendered it safe for the observer to work inside the building close to the apparatus, which greatly facilitated the readings.

EXPERIMENTS WITH WATER

Curve 9 is an isothermal curve obtained for the compression of distilled water at 4° C. plotted to a base of pressure in atmospheres, the vertical ordinates giving the volumes of a kilogramme in cubic centimetres. In the case of water the adiabatic compression was found to be only about 3 per cent. less than the isothermal. With other substances, as will be seen below, the difference was greater, on account of lower specific heat and apparently higher internal forces.

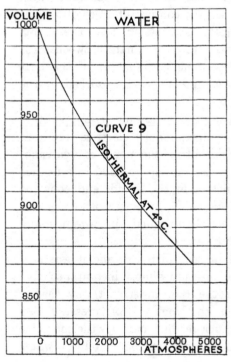
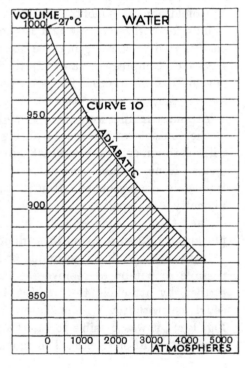

Curve 10 is a curve of adiabatic expansion of water from a temperature of 40° C. at 4550 atmospheres pressure. The temperature at the end of this process of adiabatic expansion was found to be 27° C., so that 13 calories per kilogramme have to be restored to bring the water back to 40° C. at atmospheric pressure. The external work done during expansion is given by the shaded area under the curve and amounts to 2700 kilogramme metres per kilogramme. But the 13 calories of heat which have to be added per kilogramme are equivalent to 5500 kilogramme metres, so that in passing from the initial state of 40° C. temperature and 4550 atmospheres pressure to a final state at the same temperature and atmospheric pressure 2800 kilogramme metres of energy have become latent, or in other words have been converted into internal potential energy. The increase of volume between these two conditions is 130 c.c., and this disappearance

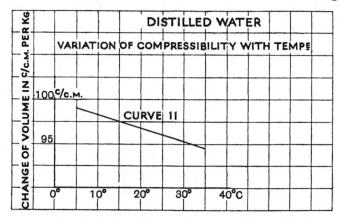

of energy is therefore equivalent to the work that would be done against an internal force of average value = 280,000 ÷ 130, or 2150 atmospheres.

It will be seen that by increasing the pressure from 1 up to 4500 atmospheres water at 4° C. is compressed to 87 per cent. of its original volume. At 3000 atmospheres it is 90·2 per cent. This agrees closely with 89·8 per cent. given by Amagat for 3000 atmospheres pressure at 0° C.

An endeavour was also made in the case of water to find the influence of temperature upon compressibility. For this purpose the water was compressed adiabatically at various temperatures. Readings were taken of the compression at 600 atmospheres and immediately afterwards at 4500 atmospheres, the interval elapsing being only a few seconds. The amount of the compression between these pressures is plotted in Curve 11, the units adopted being change of volume in cubic centimetres per kilogramme for the vertical ordinate and temperature before compression for the horizontal base. The compressibility is reduced with increase of temperature, from which it follows that the coefficient of heat expansion at high pressures is slightly greater than at atmospheric pressure. This is in agreement with Amagat's results.

EXPERIMENTS WITH ETHER

Curve 12 shows the compressibility of pure ether at 35° C., the vertical ordinate being the volume in cubic centimetres of a quantity whose volume is 1000 c.c. at 0° C. The compression for 4000 atmospheres is from 1050 c.c. to 840 c.c., the final volume being 80 per cent. of the original, so that the compression in this case is practically double that obtained for water. The

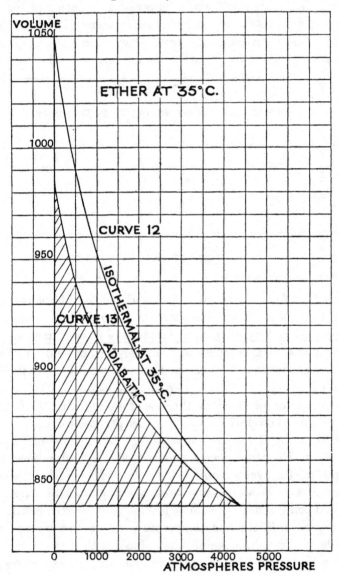

agreement with Amagat's results in this case also is fairly good, the latter giving a compression from 1050 to 863 for 3000 atmospheres at 35° C., whilst by the present experiments it appears to be from 1050 to 870, the disagreement being only about 4 per cent. of the total compression measured.

Neither for ether nor for water can the curves showing the results of these experiments be said to definitely indicate the existence of a limiting value to the compression.

In the case of ether, as in the case of water, experiments were made to ascertain the heat absorbed through the action of intermolecular forces by sudden expansion. Curve 13 is the curve of adiabatic expansion after cooling at 4400 atmospheres to 35° C. The final temperature was − 11° C. This drop of 46° C., assuming the specific heat of ether to be 0·516, corresponds to 7200 kilogramme metres of energy for unit of volume 1000 c.c. at 0° C. The external work done during expansion is given by the shaded area under Curve 13, and is 2070 kilogramme metres, leaving for the internal work 5130 kilogramme metres, representing an average molecular pressure of 2440 atmospheres.

Experiments with Paraffin Oil

Curve 14 gives the isothermal compression at 34° C. of a paraffin oil of flash point 131° F., and of specific gravity 0·812 at 0° C., and 0·788 at 34° C., the vertical ordinate being the volume of 812 grammes of the liquid. Curve 15 is the curve of adiabatic expansion after cooling to 34° C. at 4600 atmospheres. With 4500 atmospheres the oil is compressed isothermally to 871 c.c., or about 84½ per cent. of its original volume at 34° C.; the compressibility is therefore greater than for water by about 20 per cent.

The drop in temperature on expansion from 4600 atmospheres to atmospheric pressure was found to be 39° C. Taking the specific heat at 0·5, the energy extracted per 1000 c.c. is made up of 2055 kilogramme metres of external work and 4645 kilogramme metres of work against internal forces, the latter corresponding to a pressure of 2920 atmospheres.

Conclusion

To exhibit the comparison between them, the curves of isothermal compression have been replotted in Diagram II, reduced to a common basis of 1000 c.c. original volume.

Expressing compressibility in atmospheric units, that is to say, as the ratio of the decrease of volume per atmosphere of pressure to the volume of the liquid, from the foregoing experiments the following values have been deduced for the isothermal coefficients of compressibility of water, ether, and paraffin oil:

			Temperature	Pressure	Compressibility
Water	...	...	4° C.	Atmospheric	50×10^{-6}
				2000 atmos.	25×10^{-6}
				4500 ,,	$22 \cdot 5 \times 10^{-6}$
Ether	...	...	35° C.	Atmospheric	165×10^{-6}
				1000 atmos.	64×10^{-6}
				2000 ,,	$42 \cdot 5 \times 10^{-6}$
				4500 ,,	18×10^{-6}
Paraffin oil ...	...	...	34° C.	Atmospheric	87×10^{-6}
				2000 atmos.	34×10^{-6}
				4500 ,,	17×10^{-6}

The average values of the molecular force deduced from the heat lost during adiabatic expansion as described above are as follows:

							Molecular force atmos.
Water (for pressures between 0 and 4550 atmos.) ...					...	...	2150
Ether	,,	,,	0 ,,	4000 ,,	...	...	2440
Paraffin	,,	,,	0 ,,	4600 ,,	...	...	2920

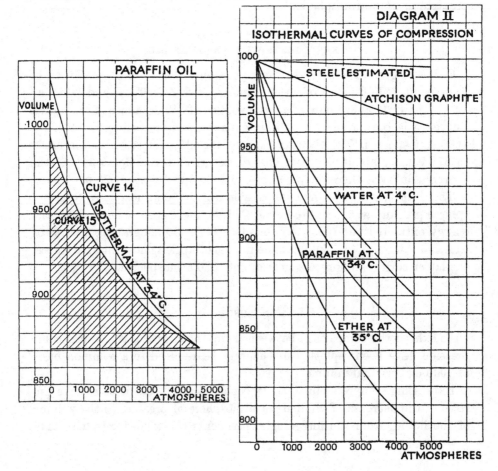

Appendix (June 3)

The method adopted in the paper for determining the equivalent internal pressure corresponding to the excess of heat extracted over the external work done is based on the following considerations:

Suppose a closed cycle on a pressure volume diagram, made up of the following processes in succession: (a) isothermal compression at temperature T_1; (b) adiabatic expansion to atmospheric pressure and temperature T_2; (c) restoration to temperature T_1 by addition of heat at atmospheric pressure.

The external work during process (c) is negligible. Hence the excess of the heat given out over external work done during (a) can be measured by the excess of the heat absorbed during (c) over the external work done during (b).

The excess thus determined in (a) has been expressed as the product of an average internal pressure into the change of volume occurring during the isothermal compression (a).

The rate of rise of temperature of fluids under compression can be determined from the thermodynamic relations resulting from the first and second laws of thermodynamics. When there is no emission of heat, we have, with the usual notation,

$$0 = K_p d\tau - \frac{\tau}{J}\left(\frac{\partial v}{\partial \tau}\right)_p dp, \tag{1}$$

from which

$$\frac{d\tau}{dp} = \tau \left(\frac{\partial v}{\partial \tau}\right)_p \Big/ J K_p. \tag{2}$$

Using α to denote the coefficient of expansion at constant pressure, or $\frac{1}{v}\left(\frac{\partial v}{\partial \tau}\right)_p$, we have

$$\frac{d\tau}{dp} = \frac{\tau v \alpha}{J K_p}. \tag{3}$$

This formula will ordinarily only bear application in the neighbourhood of atmospheric pressure, the values of α and K_p under the high pressures employed in the foregoing experiments being unknown.

Thus, for the adiabatic curve plotted for water, in the neighbourhood of atmospheric pressure, we have for the various expressions in equation (3):

$$\tau = 273 + 27 = 300, \qquad J = 42,400 \text{ kgrm.-cm. per kgrm.}$$
$$v = 1000 \text{ c.c. per kgrm.} \qquad \alpha = 0.00025,$$

from which there results $d\tau/dp = 0.00177$, or a rise of $0.00177°$ C. per atmosphere of pressure. The rise in temperature will, however, in the case of water, proceed more rapidly in the later stages of the compression as higher pressures are reached, since α increases both with the temperature and with the pressure.

With the help of Curve 11 it is possible to obtain an approximate estimate of the value of α of the fluid when under compression. Denoting by β the compressibility at constant temperature, since

$$\beta = -\frac{1}{v}\left(\frac{\partial v}{\partial p}\right)_\tau \quad \text{and} \quad \alpha = \frac{1}{v}\left(\frac{\partial v}{\partial \tau}\right)_p,$$

it follows that
$$\left(\frac{\partial \alpha}{\partial p}\right)_\tau = -\left(\frac{\partial \beta}{\partial \tau}\right)_p.$$

The slope of Curve 11 gives for the value of $-\partial\beta/\partial\tau$, with the unit adopted for β (viz. change of volume per unit volume per atmosphere of pressure), $0\cdot035 \times 10^{-6}$, and if we assume this as the approximate rate of increase of α throughout the range of pressures covered by the adiabatic compression from 27° C. at atmospheric pressure to 40° C. at 4550 atmospheres pressure, we have for the value of α at the beginning of the compression $0\cdot00025$, and at the end $0\cdot00061$, or an average value of $0\cdot00043$.

The expressions in equation (3) then assume the following average values over the whole range of this compression:

$$\tau = 306\cdot5, \qquad \alpha = 0\cdot00043,$$
$$v = 940 \text{ c.c. per kgrm.,}$$

giving $d\tau/dp = 0\cdot002925°$ per atmosphere.

The measured drop in temperature during the reverse process of adiabatic expansion, from a pressure of 4550 atmospheres at 40° C. to atmospheric pressure, was 13° C., or $0\cdot00286°$ per atmosphere, thus agreeing even closely with the preceding estimate.

THE STEAM TURBINE

The Rede Lecture delivered in Cambridge, June 8th, 1911

In modern times the progress of science has been phenomenally rapid. The old methods of research have given place to new. The almost infinite complexity of things has been recognised and methods, based on a co-ordination of data derived from the accurate observation and tabulation of facts, have proved most successful in unravelling the secrets of Nature; and in this connection I cannot but allude to the work at the Cavendish Laboratory and also to that at the Engineering Laboratory in Cambridge, and to the association of Professor Ewing with the early establishment of records in steam consumption by the turbine.

In the practical sphere of engineering the same systematic research is now followed, and the old rule of thumb methods have been discarded. The discoveries and data made and tabulated by physicists, chemists, and metallurgists, are eagerly sought by the engineer, and as far as possible utilised by him in his designs. In many of the best equipped works, also, a large amount of experimental research, directly bearing on the business, is carried on by the staff.

The subject of our lecture to-day is the Steam Turbine, and it may be interesting to mention that the work was initially begun because calculation showed that, from the known data, a successful steam turbine ought to be capable of construction. The practical development of this engine was thus commenced chiefly on the basis of the data of physicists, and, as giving some idea of the work involved in the investigation of the problem of marine propulsion by turbines, I may say that about £24,000 was spent before an order was received. Had the system been a failure or unsatisfactory, nearly the whole of this sum would have been lost.

Further, in order to prove the advantage of mechanical gearing of turbines in mercantile and war vessels, about £20,000 has been recently expended, and considerable financial risks have been undertaken in relation to the first contracts.

The first turbine of which there is any record was made by Hero of Alexandria, 2000 years ago, and it is probably obvious to most persons that some power can be obtained from a jet of steam either by the reaction of the jet itself, like a rocket, or by its impact on some kind of paddle wheel.

In the year 1888 Dr De Laval of Stockholm undertook the problem with a considerable measure of success. He caused the steam to issue from a trumpet-shaped jet, so that the energy of expansion might be utilised in giving velocity to the steam. Recent experiments have shown that in such jets about 80 per cent. of the whole of the available energy in the steam is converted into kinetic energy of velocity in a straight line, the velocity attained into a vacuum being about 4000 feet per second.

In 1884, or four years previously, I dealt with the turbine problem in a different way. It seemed to me that moderate surface velocities and speeds of rotation were essential if the turbine motor was to receive general acceptance as a prime mover. I therefore decided to split up the fall in pressure of the steam into small fractional expansions over a large number of turbines in series, so that the velocity of the steam nowhere should be great. This principle of compounding turbines in series is now universally used in all except very small engines, where economy in steam is of secondary importance. The arrangement of small falls in pressure at each turbine also appeared to me to be surer to give a high efficiency, because the steam flowed practically in a non-expansive manner through each individual turbine, and consequently in an analogous way to water in hydraulic turbines, whose high efficiency at that date had been proved by accurate tests.

I was also anxious to avoid the well-known cutting action on metal of steam at high velocity.

The close analogy between the laws for the flow of steam and water under small differences of pressure has been confirmed by experiment, and the usual formula, velocity $= \sqrt{2gh}$, where h is the hydraulic head, gives the velocity of issue from a jet for steam with small heads and also for water. I shall presently follow this part of the subject further in dealing with the design of turbines. Having decided on the compound principle, it was necessary to commence with small units; and thus, notwithstanding the compounding, the speed of revolutions though much reduced was still rather high.

I have said that steam behaves almost like an incompressible fluid in each turbine of the series, but because of its elasticity its volume gradually increases with the succession of small falls of pressure, and the succeeding turbines consequently are made larger and larger. This enlargement is secured in three ways: (1) by increasing the height of blade, (2) by increasing the diameter of the succeeding drums, and (3) by altering the angles and openings between the blades. All three methods are generally adopted to accommodate the expanding volume of the steam which in a condensing turbine reaches one hundredfold or more before it issues from the last blades to the condenser.

Now as to the best speed of the blades, it will be easily seen that in order to obtain as much power as possible from a given quantity of steam, each row must work under appropriate conditions. This has been found by experiment to require that the velocity of the blades relatively to the guide blades shall be from one-half to three-quarters of the velocity of the steam passing through them, or more accurately equal to one-half to three-quarters of the velocity of issue from rest due to the drop of pressure in the guiding or moving blades, for in the usual reaction turbine the guiding are identical with the moving blades.

The curve for efficiency in relation to the velocity ratio has a fairly flat top, so that the speed of the turbine may be varied considerably about that for maximum efficiency without materially affecting the result.

In compound land turbines the efficiency of the initial rows is about 60 per cent., and of the later rows 75–85 per cent., and considering the whole turbine, approximately 75 per cent. of the energy in the steam is delivered on to the shaft. The expansion curve of the steam lies between the adiabatic and isothermal curves, but nearer the former, because 75 per cent. is converted into work on the shaft and only 25 per cent. is lost by friction and eddies in the steam and therefore converted into heat.

In turbine design the expression of the velocity ratio between the steam and blades may be represented by the integral of the square of the velocity of each row through the turbine, and if, for instance, this integral is numerically equal to 150,000—a usual allowance for land turbines—then we know that, with a boiler pressure of 200 lb. and a good vacuum, the velocity of the blades will be a little over one-half that of the steam, and the turbine will be working close up to that speed which gives the maximum efficiency. In large marine turbines where weight and space are of importance the integral may be from 80,000 to 120,000 or more. With the first figure a loss of efficiency of about 10 per cent. below the highest attainable is accepted, and with the latter figure the deficit is only about 3 per cent.

There are many forms of turbines on the market. It is only necessary, however, for us here to consider the four chief types which are:

First, the compound reaction turbine with which we have been dealing, representing over 90 per cent. of all marine turbines in use in the world, and about half the land turbines driving dynamos.

Second, the De Laval, which is only used for small powers.

Third, the "multiple-impulse, compounded" or Curtis, which has been chiefly used on land, but which has been fitted in a few ships.

Lastly, a combination of the compound-reaction type with one or more "multiple-impulse or Curtis elements" at the high-pressure end to replace the reaction blading.

We may dismiss the other varieties as simply modifications of the original types without possessing any originality or scientific interest.

Now let me further explain the multiple-impulse type, and commence by saying that it is the only substantial innovation in turbine practice since the compound-reaction and the De Laval turbines came into use. It was proposed by Pilbrow in 1842, and first brought into successful operation by Curtis in 1896. Some consideration should be given to it as involving several characteristic points of difference from what has been said about the compound-reaction type. Curtis, in the first place, uses the De Laval divergent nozzle, and he also uses compounding to the limited extent of only 5 to 9 stages, as compared with 50 to 100 in the compound type. With these provisos the same principles in the abstract as regards velocity ratio now apply, and the steam issuing from the jets rebounds again and again between the fixed and moving buckets at each velocity-compounded stage: the best velocity ratio in a 4-row multiple impulse is only one-seventh and the efficiency about 44 per cent., and therefore much lower than that of reaction blading, which as we have stated is under favourable conditions 75–85 per cent.

The advantages, however, to be derived from the use of some multiple impulse elements at the commencement of the turbine are that because there is very little loss in them from leakage, therefore in spite of their low intrinsic efficiency, one or more multiple impulse wheels can in certain cases usefully replace reaction blading. The explanation is that in turbines of the compound-reaction type of moderate power and slow speed of revolution the blades are often very short at the commencement, and consequently there is in such cases excessive loss by leakage through the clearance space, which brings the efficiency below that of impulse blading. In most cases one multiple-impulse wheel is preferred, followed by reaction blading.

When highly superheated steam is used the temperature is much reduced by expansion in the jets and work done in the impulse wheel before it passes to the main turbine casing.

The highest efficiency yet attained by land turbines has, however, been with the pure compound-reaction type of large size, where the high pressure portion is contained in a separate casing of short length and great rigidity, now made usually of steel. The working clearances can by this arrangement be reduced to a minimum and the highest efficiency attained.

Table I. *Performance of Parsons' turbo-generators at different epochs.*

Date	Power (kw.)	Steam per kw. hour (lb.)	Vacuum (Bar. 30″) (in.)	Superheat (° F.)	Steam pressure per sq. in. (lb.)
1885	4	200	0*	0*	60
1888	75	55	0*	0*	100
1892	100	27·00	27	50	100
1900	1250	18·22	28·4	125	130
1902	3000	14·74	27	235	138
1907–10	5000	13·2	28·8	120	200

* These were non-condensing turbines using saturated steam.

Table II. *Performance of notable ships of different epochs with Parsons' turbines*

Date	Name of ship	Length (ft.)	Displace-ment (tons)	Horse-power	Steam consump-tion per s.h.p. per hour for all purposes (lb.)	Speed in knots
1897	*Turbinia*	100	44½	2,300	15	32·75
1901	*King Edward*	250	650	3,500	16	20·48
1905	H.M.S. *Amethyst*	360	3,000	14,000	13·6	23·63
1906	H.M.S. *Dread-nought*	490	17,900	24,712	15·3	21·25
1907	*Mauretania* and *Lusitania*	785	40,000	74,000	14·4	26·0

In conclusion, I would venture to predict that the use of the land and marine turbine will steadily increase, and that the improvements that are being made to still further increase its economy will for a long time enable the turbine to maintain its present leading position as a prime mover.

THE MARINE STEAM TURBINE FROM 1894 TO 1910

Institution of Naval Architects, Jubilee Meetings, July 5th, 1911

It was very early realised that the suitability of the turbine for steam of very low pressure, which had been proved by the high percentage of power realised in the low-pressure portion of land turbines, would find an important application in ships by working the turbine from the exhaust of reciprocating engines, but not until 1901 was the first combination vessel, the *Velox*, a 30-knot destroyer, laid down, as a speculation, by the Parsons Marine Steam Turbine Company, Ltd. Her main propelling machinery was nearly a duplicate of the *Viper's*, but, in order to increase the economy at speeds below 13 knots, two triple-expansion engines of 150 horse-power each were coupled to the low-pressure turbine shafts through flexible and detachable clutch couplings. At low speeds the reciprocators exhausted into the high-pressure turbines; at speeds above 13 knots the engines were uncoupled. The vessel was acquired by the Admiralty in 1903. The same year H.M.S. *Eden* of practically the same dimensions as the *Velox* was launched, and cruising turbines in series were fitted in her instead of reciprocating engines.

In 1902 the Admiralty placed the order for the third-class cruiser *Amethyst* with turbines, and for three sister vessels with reciprocating engines of 10,000 horse-power. The turbine installation of the *Amethyst* was of the usual three-shaft arrangement, the high-pressure turbine driving the centre propeller and the two low-pressure turbines in parallel on the steam driving the wing propellers, a high-pressure cruising turbine and an intermediate-pressure cruising turbine being also directly coupled through flexible claw couplings to the low-pressure turbine shafts; these latter were in series on the steam with the main high-pressure turbine. The trials of these vessels conclusively proved the superiority of the turbine in water consumption.

Subsequent to the official trials arrangements were made for utilising the auxiliary exhaust in the turbines, when a further increase in economy was obtained.

The comparative trials of these vessels had a great influence upon the future of the turbine.

As a general rule, the larger and faster the vessel, the more easy it has been to arrive at a satisfactory and profitable all-turbine solution as regards efficiency and first cost. Below the sea speed of 16 knots, the solution is not altogether favourable, and very few such vessels have been fitted. A slight lowering of the boundary of suitable speed exists in very large vessels, and also may be effected by the employment of the impulse principle at the high-pressure end, as more fully explained later.

This most important question of applying the turbine to lower speed vessels has from the commencement received consideration, and one satisfactory solution appeared to lie in the Combination System to which we shall further allude. Another solution with a somewhat different scope has more recently been investigated; it is the use of double helical gearing of the De Laval class on a large scale, for gearing the high-pressure portion only, or in some cases the whole of the turbines on to the screw shafting. Both of these solutions have given satisfactory results, but the reduction gearing appears to be the more important as applicable to vessels of all speeds. Both systems are described more fully later.

The first turbine battleship *Dreadnought* was laid down at Portsmouth in October, 1905. The primary contract for the whole of the machinery was placed with Vickers, Sons and Maxim, Ltd., the Parsons Marine Steam Turbine Company, Ltd. being sub-contractors for the turbines. She has four shafts driven by two high-pressure and two low-pressure turbines, whilst two cruising turbines are coupled to the low-pressure turbine shafts. The astern turbines consist of a high-pressure astern in a separate casing on the high-pressure ahead shaft, in series with a low-pressure astern incorporated in the main low-pressure turbine casing. The cruising turbines are in parallel and not in series on the steam as in the *Amethyst*; the high-pressure, low-pressure, cruising, and astern turbines comprising one propelling unit on each side of the vessel.

At full power the steam consumption of the *Dreadnought* was 13·48 lb. per shaft horse-power per hour, while in the succeeding battleships of the class it averaged 13·01 lb., and in the three cruisers of the *Invincible* class 12·03 lb. With reciprocating engines nearly 16 lb. would be a fair average, and it thus follows that a great reduction in boiler weights was permissible. Again, the high efficiency of the low-pressure turbine made it well worth while to pass the exhaust steam from the auxiliary engines to this turbine instead of to the condenser. Indeed, the exhaust steam in some battleships has been proved to be alone sufficient to drive the vessel at a speed of 5 to 6 knots.

The coal consumption at full power of the three 26-knot armoured cruisers of the *Invincible* class ranged from 1·2 lb. to 1·7 lb. per shaft horse-power per hour, the average for the three ships being 1·47 lb. per shaft horse-power. In the three cruisers of the *Minotaur* class, with piston engines, it was 1·8 lb., and in the six cruisers of the *Duke of Edinburgh* or *Warrior* class 2·1 lb. per indicated horse-power per hour. On the thirty hours' endurance trial at 70 per cent. of the total power, the turbines also proved more efficient, although the advantage was not so marked.

At one-fifth power the coal consumption of the three *Invincible* cruisers averaged 2·4 lb. per shaft horse-power per hour, as compared with 1·87 lb. per indicated horse-power per hour in the *Minotaurs* and 2·05 lb. in the *Duke of Edinburgh* cruisers.

In the mercantile marine, the first vessel to be fitted with turbines was the Clyde passenger steamer, *King Edward*, built to the joint ownership of Captain John Williamson, Messrs Denny, of Dumbarton, and of the Parsons Marine Steam Turbine Company, Ltd. Her length is 250 feet, and with 3500 horse-power she attained a speed of 20·48 knots. Her success led to the construction of a second vessel for the Clyde passenger traffic in 1903, and in the same year the *Queen* was built for the Dover and Calais route. All these vessels have three shafts, the high-power turbine in the centre exhausting into two low-pressure turbines on the wing shafts.

Thus by 1904 two of the most suitable fields for the marine turbine had been entered, namely, for vessels of war and cross-channel and passenger service. By 1905 the turbine was being adopted for nearly all new cross-channel steamers of high speed built in this country, and about a year later by the British Admiralty for all new construction.

The application to large liners remained as yet untouched. The first vessels to be fitted with turbines for Transatlantic service were ordered by the Allan Line, viz., the *Victorian* and *Virginian*. This marked a notable step in advance in the application of the new system to ocean-going ships of high speed.

At this juncture the late Lord Inverclyde appointed a commission of experts to investigate the suitability of the turbine for two express Cunarders for the New York route. After most careful consideration of all data then available, and in view of much additional experimental research conducted by the Committee, as well as tests on large land turbines, and on existing turbine vessels, the Committee unanimously recommended turbines in preference to reciprocating engines for the *Mauretania* and *Lusitania* of 70,000 horse-power and 24½ knots sea speed.* The performance of these vessels has justified the decision; a mean speed of 26 knots has been maintained in favourable weather across the Atlantic, and an average speed of 25½ knots has been maintained on many successive voyages. This step completed the entry of the turbine into all classes of fast vessels for which it was at the time deemed suitable, and its adoption for fast vessels has since been almost universal.

In 1904 the general policy of granting licences on easy terms was decided

---

* The *Lusitania* was sunk by German torpedoes off the Old Head of Kinsale, with heavy loss of civilian life, in May 1915. After the War the *Mauretania* was reconditioned and equipped for oil firing, and in August 1929 made an average speed of 27·22 knots between the Ambrose Channel and Plymouth.

upon by the Parsons Marine Steam Turbine Company, Ltd., and by the end of 1905 a large number of shipbuilding and engineering firms had acquired marine licences, the number having considerably increased up to the end of 1910. It is now clear that the broad policy then adopted has been conducive to the exceptionally rapid adoption of the turbine.

The combination system

The *Velox*, in 1901, as has been stated, was the first combination vessel. The system was described and advocated by me in my Presidential Address to the Institute of Marine Engineers in January, 1905. The *Otaki*, of 464 feet length, 9900 tons deadweight capacity, built and engined by Messrs Denny, of Dumbarton, in 1908, for the New Zealand Shipping Company, was the second; the sister vessels, the *Orari* and *Opawa*, were fitted with triple-expansion engines.

About the same time the *Laurentic* and *Megantic* were laid down by Messrs Harland and Wolff for the Canadian service. They are 565 feet long and about 20,000 tons displacement. The first was fitted with triple-expansion engines driving the side screws and a low-pressure turbine driving the centre screw. The latter vessel was fitted with the usual quadruple-expansion engines and twin screws. In service it is stated that the combination vessel consumes 12 to 14 per cent. less coal than the sister vessel.

In 1909 Messrs Harland and Wolff commenced the *Olympic* and *Titanic*, 882 ft. 9 in. long and about 60,000 tons displacement, the arrangement of combination machinery being similar to that of the *Laurentic*; with a collective horse-power of 30,000 the designed sea speed is 21 knots. The same firm is fitting the combination system in the *Demosthenes*, of 19,500 tons displacement, to steam 13 knots, being built for Messrs George Thompson and Co., Ltd., and also in a large ship for the Royal Mail Steam Packet Company's service to the South American republics.

The Orient Line ordered combination machinery for the *Orama*, laid down in 1910, for their Australian mail service. The vessel is 550 feet long, and of about 18,000 tons displacement, and is being constructed and engined by Messrs John Brown and Co., Ltd., Clydebank.

All these ships have triple screws, each reciprocating engine driving a wing shaft and propeller, while the turbine, taking the exhaust steam from both engines, works the centre screw. This arrangement has the advantage that, in manœuvring and for going astern, the piston engines only are used, and thus there is no need for an astern turbine.

At St Nazaire, in France, an intermediate liner, the *Rochambeau*, is building, and she has four shafts, the reciprocating engines driving the inner shafts and the turbines the wing propellers.

8-2

REDUCTION GEARING BETWEEN THE TURBINE AND THE PROPELLER

Several forms of gearing have been proposed. Electrical gearing was fitted, I believe, by the Heilmann Company in a small vessel about thirteen years ago, but little is known of the results.

Continuous current generators and motors were originally proposed, but of late years some form of alternating current transmission has been preferred by the advocates of the system.

Within the last few months a 50-feet vessel has been fitted by Mr Henry Mavor, of Glasgow, with his form of alternating current generator and motor, and a suction gas engine directly coupled to the generator. A quick reversal or graduation of speed of the propellers can be effected without altering the speed of the engine by this method. The efficiency of the electrical transmission is said to reach 88 per cent.

The same means of transmission is applicable to the steam turbine.

A hydraulic transmission gear has been developed by Mr Föttinger, and has been fitted in two small vessels. The efficiency of transmission is said to reach 86 per cent.

The first application of geared turbines was made by the Parsons Marine Steam Turbine Company, Ltd., in a 22 feet launch in 1897.

Double helical reduction gear between the turbine and the propeller has so recently been the subject of papers before this Institution, that it is unnecessary to say more than that it promises to be an important factor in many marine turbine installations in the future. With an efficiency of over 98 per cent., and a remarkable absence of wear, it will add considerably to the efficiency of turbine installations even in fast vessels, and, what is of greater importance, it makes the turbine applicable to all classes of vessels.

The London and South-Western Railway Company have recently decided to fit two new steamers of 5000 horse-power and 20 knots speed with geared turbines. They will have twin screws and a high-pressure and a low-pressure turbine will be geared on to each screw shaft similarly to the arrangement in the *Vespasian*.

Although the manufacture of land turbines began to be general on the Continent in 1900, it was not till 1902 that the first turbine-propelled vessel was ordered by the French Admiralty, a torpedo boat, 130 feet long, 94 tons displacement, and 26½ knots speed. She had three shafts, and the turbines, which were built by the Parsons Marine Steam Turbine Company, Ltd., at Wallsend-on-Tyne, were in series as in the *Turbinia*, but with one cruising turbine coupled to the low-pressure turbine which drove the centre shaft, and with a reversing turbine in the low-pressure casing.

The first marine turbines fitted in a merchant steamship built in France were made in 1908 at the works of Cie Electro-Mécanique for the Mediterranean steamer *Charles Roux*, of 9000 horse-power and 19 knots speed.

In Germany the marine turbine was first introduced into the torpedo-boat destroyer *S.* 125, ordered towards the end of 1902. This vessel was one of the class of six boats ordered that year, the remaining boats having the ordinary type of reciprocating engines. In the following year the cruiser *Lübeck* was ordered by the German Admiralty to be fitted with Parsons turbines. A series of comparative trials was made with this vessel and a sister vessel, the *Hamburg*, fitted with reciprocating engines. The results obtained on service of the destroyer *S.* 125 and the cruiser *Lübeck*, as compared with those of the sister vessels fitted with reciprocating engines, were conclusive in favour of the turbine boats.

The first vessel to be fitted with Parsons turbines in the United States was the passenger steamer *Governor Cobb*. This vessel, built by the W. and A. Fletcher Company in 1905, is 290 feet long, 2900 tons displacement, and 20·5 knots speed. The arrangement of machinery is identical with that adopted in the *King Edward*.

In Japan four merchant vessels were fitted with Parsons turbines in 1908. Two of these, the *Tenyo Maru* and *Chiyo Maru*, were built at the Mitsu Bishi Dockyard at Nagasaki, the turbines for which were constructed by the Parsons Marine Steam Turbine Company, Ltd. Two others, the *Hirafu Maru* and the *Tamura Maru* were built and engined by Messrs Denny, of Dumbarton.

In Italy the cruiser *San Marco* is the first turbine ship for the Royal Italian Navy. On her trials she attained a speed of 23·7 knots, being nearly two knots in excess of the speed attained by a sister vessel, the *San Georgio*, fitted with reciprocating machinery.

All the principal navies of the world are now following the lead of the British Admiralty. In Russia four large battleships of high power are under construction. In Austria the scout *Admiral Spaun* was ordered in 1907, and on her trials she attained a speed of 27·07 knots, with 25,000 shaft horse-power. Three battleships and 12 torpedo boats are now on order in Austria with Parsons turbines, some of which are now building. Sweden and Denmark have recently ordered their first turbines for warships—a 7000 horse-power set for a 21-knot cruiser in the case of Sweden, and engines of 4000 horse-power for some twin-screw torpedo boats in the case of Denmark. In Spain three battleships, three destroyers, and ten torpedo boats, comprising the important recent naval programme of this navy, are to be fitted with turbine machinery. Portugal is having one destroyer built. Brazil has now two turbine scout cruisers, *Bahia* and *Rio Grande do Sul*, on service. These two vessels on trial averaged a speed of $27\frac{1}{4}$ knots. The Argentine Government are adopting turbines for their four new destroyers now being built by Messrs Cammell, Laird and Co. China is having three small cruisers built. Belgium does not possess a war navy, but three turbine vessels were built for the Government passenger line, the *Princesse Elisabeth*, speed 24·03 knots, in 1905, and the *Jan Breydel* and *Pieter de Coninck*, 24·3 knots. Peru possesses two passenger steamers, *Huallága* and *Ucayali*, of 6000 tons displacement and 19 knots speed. The British Colonies of Canada, Australia, and New Zealand are having various classes of war vessels fitted with the turbine system.

PRESIDENTIAL ADDRESS

North-East Coast Institution, October 25th, 1912

This summer has seen the celebration of the centenary of the *Comet*, the first really successful steam boat, and it seems fitting to recall a few of the steps that have led to the great world-wide development of the steam vessel during the hundred years that have elapsed, and which have so greatly changed and improved the conditions of life and increased the powers of production of human effort.

In all developments of machinery and shipbuilding, invention bears its part, and it is interesting to recall what Lord Fletcher Moulton once said when he was discussing the merits of a patent. He had seldom or never known of an invention which appeared in the full panoply of perfection, but it was generally to be likened to a young sapling in a great forest of large trees. It struggles for existence among them, they deprive it of the sun, their roots rob it of nutriment, and its chances of growing up and taking its place among the great trees of the forest is remote, unless perchance some unlikely circumstance may assist it to gather strength. So an invention by itself has generally little chance unless it is suitable to the times and assisted by some friendly hand in the shape of brains and capital.

There is no doubt that in these days, useful inventions come principally from trained brains, and the number of brains that have passed through schools, advance schools, colleges, and universities increases yearly, and concurrently every year the forest becomes denser; consequently we have to dig deeper and training and specialisation have become more necessary for success in any line. One hundred years ago it was possible for one brain to master many subjects; now the range and complexity is so great that it has become imperative to specialise if more than a moderately superficial knowledge is desired.

The application of steam for the propulsion of vessels had occurred to many persons before the start of the *Comet* in 1812. Roger Bacon, in the thirteenth century, anticipated that "Engines of navigation might be made without oarsmen, so that the greatest river and sea ships, with only one man to steer them, may sail swifter than if they were fully manned."

It was in the minds of De Garay about 1545, and Papin and Savery about 1695, when the Marquis of Worcester was working at steam propulsion for land carriages; Papin appears to have first introduced the steam piston, and the Marquis of Youffrey d'Abbas experimented on the river Doubs with a steam boat about 1790, while Dr John Allan in 1730 proposed jet propul-

sion, and in 1736 Jonathan Hulls obtained a patent for driving paddle
wheels by a kind of rope gear from a Newcomen engine. But the nearest
approach to success seems to have been the pleasure boat owned by Miller,
engined by Symington in 1788, and the *Charlotte Dundas*, engined by
Symington in the following year.

The American engineer, Fulton, who was on board Symington's steam
boat in 1801, after spending some time in experiments on the Seine, and
making inquiries of several engine builders, gave an order to Boulton and
Watt in 1803 for an engine of 19 horse-power, which was shipped to
America early in 1805, and in an American hull made her first voyage on
the Hudson river in 1807, when a speed of nearly four miles per hour was
attained. Watt himself was then 71 years of age, and had retired from
active business.

The *Comet*, however, in 1812, associated with the name of Henry Bell,
was the first real practical success in British waters. She commenced to ply
regularly between Glasgow and Greenock in August of that year, and soon
Clyde steamers became famous, and caused steam to be rapidly adopted.

The invention of James Watt was undoubtedly the chief cause of the
success of the *Fulton* and also of the *Comet* where others had failed, for the
invention of Watt meant a reduction in one step of more than 6 to 1 in coal
consumption over any previous engine, and we can therefore understand
why the earlier attempts had proved non-paying speculations and failures.

For many years Watt had been interested in the subject of steam navigation, but too much engaged with the perfecting of his pumping engine to
spare time for the application of his great invention of the separate condenser to marine purposes, but in 1770 he appears to have had clearly in his
mind the screw propeller.

The laws of steam which Watt discovered are simply these, that the
latent heat is nearly constant for different pressures within the ranges used
in steam engines, and that consequently the greater the range of expansion
the greater will be the work obtained from a given amount of steam or of
coal, and he also showed that the expansive force of the steam continues
down to the vacuum in the condenser. Watt had formed a clear idea as to
the advantages to be derived from the compounding of two or more
cylinders in series, but in his day materials and workmanship did not admit
of high pressures, and its realisation remained in abeyance until engineering
and metallurgy had so far advanced that boilers suitable for high pressures
could be constructed. It is clear, therefore, that in 1800 it had been
established by Watt's investigations that a high ratio of initial to exhaust
pressure was essential to economy. Watt worked at the low pressure end of
the expansion, and developed this as far as practicable, using a single
cylinder because steam pressures being then low, frictional losses and cost

put compounding out of court. The adoption of higher pressures came gradually, spurred on chiefly in the attempts at road locomotion, and later by the introduction of the locomotive, it being essential for such purposes that space, weight and consumption should be reduced to a minimum. When higher steam pressures were first attempted on board ship, there were troubles from salt in the boilers, inseparable from jet condensation; the jet condenser, however, was superseded by the surface condenser between 1840 and 1860, and then pressures rapidly rose and the compound, the triple and the quadruple expansion engine became the leading types, and have remained so with slight modifications in details.

It indeed seems that the reciprocating engine is not now capable of much further improvement. It is in truth the most efficient machine for dealing with steam of high densities but not for low densities, because with low steam pressures on the piston frictional losses absorb most of the power, and also when it is attempted to provide sufficient cylinder capacity to carry the expansion down to low pressures, the engine becomes too bulky, expensive and fragile. For these reasons a 7 lb. release pressure in the low pressure cylinder and an expansion ratio of 16 has been about the limit practically attainable.

Of late years the turbine has been applied to carry the expansion further, either alone or in combination with the reciprocating engine whose exhaust may pass to the turbine at say 10 lb. absolute, and the expansion be completed to say 1 lb. absolute; a combination system which has given very high economy, and has been fitted in the *Otaki, Laurentic, Titanic, Olympic, Rochambeau* and some other vessels, resulting in a saving of 10 to 14 per cent. over quadruple engines. The "combination" system is also adopted in two large vessels building by Messrs Denny and Messrs Swan, Hunter and Wigham Richardson, Ltd. The turbine alone has superseded the reciprocating engine in fast vessels, chiefly because of its superior economy in fuel, the direct result of its greater range of expansion, or, in other words, its power of utilising the expansive energy of steam right down to the pressure in the condenser, and fulfilling more completely the laws discovered by Watt.

The total horse-power of the world's shipping has been recently estimated at about 26,000,000, and of this about 8,500,000 is now provided by turbines.

The turbine is, however, ill adapted for direct coupling to the screws of slow ships, but the application of gearing, of whatever type, removes this embargo, and by its use both the propeller and the turbine are allowed to run at such speeds of revolution as are most suitable to give the greatest propulsive horse-power in proportion to the steam used. The three principal types of reduction gear at present before us are the hydraulic, the electric, and the helically cut mechanical gear. Which of these is preferable depends on considerations of efficiency, first cost, weight, durability and reliability.

The efficiency ascertained for the hydraulic is 86 per cent. at full power in moderate sizes and 92 per cent. anticipated in large sizes; for the electric 92 per cent. has been anticipated, but, so far as I am aware, has not been realised in practice; for the mechanical, over 98 per cent. has been realised at full power.

The hydraulic gear or transformer invented and perfected by Dr Föttinger may be described as a centrifugal pump discharging directly into a water turbine, which is concentric and co-axial with it but mounted on a separate shaft. In the annular space between the pump and the turbine, the guide blades are set to deflect the water suitably, and by their angle and shape the speed of the turbine in relation to the pump is determined and can be varied within wide limits by this means, so that the turbine may be made even to revolve in the reverse direction to the pump. The method, however, proposed by Dr Föttinger for marine reduction gear is to have one transformer for going ahead and another for going astern, which are filled or emptied alternately, according to the direction of motion desired, a supply pump, tanks and valves being provided for the purpose. The whole apparatus is small in comparison with the power transmitted. For certain purposes it promises to be very useful. Whether it will receive general adoption for marine propulsion gearing does not seem clear until further and lengthened tests have been made.*

Electrical gearing has been before the public for many years. Heilman was among the first to propose its use in 1890 as a variable gear interposed between a fast running compound engine and the wheels of a locomotive, and I believe he tried it on a small vessel before this date. Several of these engines were constructed, and in 1895 my firm were asked to quote for turbo-generators to take the place of the compound reciprocating engines and dynamos hitherto used. I was not at the time favourably impressed with the scheme because of the cost, complication, weight and risks of breakdown, and the offer was declined. The system, however, where the current is generated on the locomotive itself has not been applied beyond the first two or three examples built in France, except in a modified form for motor vehicles such as in the Thomas transmission.

Mr Mavor has recently constructed a small experimental vessel on the Clyde with electrical gearing which has given satisfactory results, and the General Electric Co. of America are at present supplying to the order of the United States Government an installation of 6000 horse-power of turbo-alternators supplying current to electric motors on the screw shafts for the propulsion of a collier of 20,000 tons at the speed of 14 knots, and the results will be received with interest.

* *Transactions North-East Coast Institution*, Vol. XXVIII, pp. 287, 309; *Transactions, Schiffbautechnische Gesellschaft*, 1910, p. 157.

Mechanical gearing commenced to be used commercially aboard ship in 1836, when screw propulsion was tried in place of paddle wheels; so accustomed were engineers to slow running paddle engines that they feared to couple the engine directly to the relatively higher speed screw, and up to about 1860 mortice gearing was in general use. It consisted of a large wheel on the engine shaft, gearing into a smaller wheel on the screw shaft. The teeth were in several rows and staggered, those of the large wheel being of hard tough wood, such as hornbeam, held in position by being driven into suitable holes in the periphery. The smaller wheel was of iron, cast with its teeth.

In the case of the P. and O. liner *Simla*, built in 1850, of 2640 tons and 640 nominal horse-power, with Steeple engines working with 17 lb. steam pressure, the ratio of the gear was 2·75 to 1, and the teeth on the gear were in four rows and staggered. A large model of this engine is in the South Kensington Museum.*

Though such gearing could only be regarded as a temporary expedient, it was fitted in a large number of ships; its continued use for many years is an interesting feature in view of the gradual but more satisfactory solution of the problem furnished by the development of the fast running direct coupled reciprocating engine. It is also of interest in view of the re-introduction of gearing, as a specific means of reconciling the inherent and, I believe, permanent idiosyncrasies of the turbine and the screw propeller.

THE SUPERHEATER

The superheater must now claim our attention. It was first used in the period of low steam pressures and jet condensation and simple engines. Under these conditions it added greatly to the economy in fuel, usually from 20 to 30 per cent., because the low steam pressures and, therefore, temperatures then in vogue permitted a large amount of superheat to be added without risk of destroying lubrication and packings, and for the further reason that it annulled condensation and re-evaporation, which is so large a factor of loss in the single expansion condensing engine. The difficulty which prevented its general adoption arose chiefly from the presence of salt in the boilers, inseparable from jet condensation which was then universal. Salt, carried over in small quantities with the steam, soon forms a layer of sodium and magnesium chloride in the superheater, accompanied by the liberation of chlorine from the latter, causing rapid corrosion. In some instances serious explosions occurred. Superheaters consequently fell into disfavour. The supersession of jet condensers by surface condensers in the 'fifties removed to a large extent this difficulty. Condensers, however, had to be kept tight and feed make-up apparatus

* [Now in the Kelvin Grove Museum, Glasgow. Ed.]

provided of sufficient capacity to supply the boilers with fresh water. At one period these conditions could not always be ensured, but of late years condenser troubles, which generally took the form of corrosion and perforation of the tubes, have been largely overcome, feed filters have been introduced, and as a consequence superheaters are now being adopted more extensively and successfully.

Sir Henry Oram, in his Presidential Address to the Junior Institution of Engineers in 1909, says that superheaters were fitted experimentally by the Admiralty during the period 1830 to 1858, and that an Admiralty Committee was appointed in 1858; superheaters were then always specified from 1863 to 1870, but as a result of trouble from corrosion and leaks, and the transference of metal and rust from pipes and superheaters into the engines, their use was discontinued, the adoption of higher steam pressures and compounding of engines being then thought to be a preferable means of securing economy.

About 600 vessels of small and moderate size, fitted chiefly with the Schmidt superheater, have now been working satisfactorily for some years, resulting in a saving of coal of between 10 and 30 per cent. The Wilson Line have thirteen ships fitted with superheaters which, I am informed, have given little or no trouble, and produced a considerable saving in coal consumption.

The present position seems to be that with reasonable care to the exclusion of salt from the boilers, well-designed superheaters are a practical success at sea, and result in a considerable and important saving of fuel.

In this connection it is interesting to note that on land superheaters with high degrees of superheat in conjunction with turbines have for some years become practically universal in all large power plants for the generation of electricity in this and other countries. The structural parts of the turbines subjected to high superheat being made of steel because at temperatures over 550° F. cast-iron gradually distorts and loses strength.

INTERNAL COMBUSTION ENGINES

Gaseous fuels are mostly produced from coal in its various forms, and are, therefore, of the first importance in this country, at least to internal combustion engines, but liquid fuels are also of great moment in view of the recent development of the petrol engine and engines using paraffin and the heavy and waste oils.

The application of the Diesel engine to large vessels such as the *Selandia*, Dr Diesel's recent enthusiastic advocacy of heavy oils at the Institution of Mechanical Engineers, and Sir Marcus Samuel's statements as to the amount and probable price of such oils have, however, tended to distort in the public mind the true relationship of coal and oil fuels. It is accordingly desirable to discuss the position briefly.

The internal combustion engine has been in use for over thirty years, during the earlier period in the form of the comparatively slow running gas engine for land use. Then the perfecting of electrical ignition, the introduction of petrol, and the demand for a light engine brought into use the modern petrol engine for motor cars and other purposes. Concurrently, the engine for burning somewhat heavier oils has been introduced, wherein the oil is gasified by being injected into a hot pot initially heated and in communication with the cylinder end. In all these engines the pressures of compression are comparatively low. In the Diesel engine, however, which can burn still heavier oils and in which the compression is much higher, the heat produced by such compression is sufficient both to gasify and to ignite the injected fuel. The total number of Diesel engines at work is, however, comparatively small, Mr Dugald Clerk stating at the meeting of the British Association this year that it did not equal one year's output of a large gas engine maker. For comparatively small powers, as viewed by the marine engineer, internal combustion engines have proved eminently satisfactory, and the economy of fuel for such sizes is very much superior to that of steam engines; chiefly for this reason, they have almost entirely superseded the latter for small powers. On the other hand, as the size increases their relative superiority diminishes until in the case of the largest steam units, as compared with the largest internal combustion units hitherto constructed, there is very little difference in the consumption of fuel between the two.

As regards weights and cost per horse-power of gas engines, a turning point is reached at about a 16-inch diameter of cylinder, beyond which the cost and weight per horse-power increase rapidly with the size, and in the Diesel this point is reached at a still smaller size of cylinder; whereas in reciprocating steam engines this turning point only becomes apparent when a very much larger size and power of cylinder is reached, such as in large modern liners, and further, in the turbine this turning point is never reached except in the largest and most powerful vessels.

I alluded this spring to some of the difficulties that arise from the intensity of the heat acting on the cylinder walls and internal parts of large internal combustion engines, and to the reasons why such adverse conditions increase with the size of the cylinder; whether some method may be discovered to reduce the heat or its deleterious effects on the engine remains to be seen. The proposal of Professor B. Hopkinson to keep the internal surfaces cool by the injection of water spray may be a solution; it is a recent proposal, and has, so far, received only experimental application, which, on the whole, has been encouraging.

At the present time the position seems to be that as regards marine work the only satisfactory means of constructing a large internal combustion unit of power for gas or oil is to multiply the cylinders; there are several ways of doing this. If the ship is sufficiently fast a very large number of cylinders may be grouped on one line of shafting, the scantlings of the cranks and shafting at the after parts being much enlarged to meet the increased torque, and between every six or eight cylinders there may be placed suitable flexible couplings to ensure alignment. Such an arrangement is, however, not suitable to slow vessels, because the dimensions of the propeller to absorb the large power necessitate a too slow rate of revolution for the engine.

A multiplicity of shafting and propellers which would suit the engineer is a solution, but would not be favoured by the naval architect.

Another and preferable method which naturally suggests itself is the use of gearing by which a large number of multiple cylinder internal combustion engines may be grouped on to and drive pinions each gearing into a large wheel to drive the screw, and there may be several large wheels on each line of shafting. In such an arrangement, each engine would have its flywheel; and a spring drive with torsional damping might be interposed between it and its pinion to reduce the irregularity of torque.

Any of these solutions appears, at first sight, complex, involving a very great multiplicity of working parts, which, however, should not be assumed as an insuperable objection. For instance, the large number of blades in a turbine was once held as a serious objection, but experience has shown that this is not the case; and with a multiplicity of internal combustion engines there undoubtedly follows increased safety from serious or total breakdown, provided that suitable means are provided for disconnecting any damaged unit and also for preventing, in case of such failure, any damage to the rest of the system.

GAS TURBINES

A few remarks about the gas turbine, of which we have heard a good deal of late, may prove interesting.

Undoubtedly the development of the steam turbine, displacing as it has done nearly all the larger reciprocating units on land and sea, and the difficulty of the large cylinder gas engine have stimulated inventors to try their hand at the gas turbine. A few experiments I made many years ago, some of which were directed to ascertain the possibility of the gas turbine, convinced me that with the metals now at our disposal, the cylinder and piston was the best way of utilising the power of gaseous explosions, and that to attempt to cause flame, whether at full heat or cooled by admixture of steam or water to impinge on blades was the wrong method of utilising the power.

I have studied the subject very closely up to the present time, and am still of the same opinion.

The large scale experiment of M. Réné Armengaud, made some six years ago, and described fully in H. H. Suplee's book on *The Gas Turbine*, reached a stage which showed that when air and gas or oil in proper proportions for complete combustion were pumped into a combustion chamber lined with carborundum, and sufficient steam admitted to reduce the temperature of the products to about 400° C. before they passed to the expansion nozzle, whence they played on a De Laval wheel, after deducting the negative work required for the pumps, the power realised was at the best only about one-half of that of a good piston internal combustion engine using the same quantity of fuel.

More recently Mr Holzwarth has constructed a still larger scale gas turbine at Mannheim of the intermittent explosion type. Ten stationary chambers are alternately charged by a turbo-compressor with gas and air in suitable proportions, and discharge the products when ignited through expanding jets on to a Curtis turbine wheel. In this gas turbine the compressor and pump are worked by steam generated by the waste heat of the exhaust gases so that no negative work is required from the gas turbine itself. The chambers are charged to about two atmospheres before explosion. In analysing the results of this turbine, Mr Dugald Clerk*, in his paper to the British Association this summer, thinks, and I agree with him, that the results claimed by the inventor are erroneous and that the probable efficiency is rather less than that obtained by Armengaud, or under one half of that of the piston engine, and that the weight of the Holzwarth machine is against its adoption for many purposes.

Experiments have also, I understand, been made by Mr Fullagar, of Newcastle, on this subject.

I am on the whole inclined to think that possibly a more hopeful course than that of working on the lines of the gas turbine, or the gas propelled water turbine, will be found in some form of rotary engine of the disguised piston type, in spite of such constructions having, after numerous attempts, been failures as steam engines. In this connexion one might possibly call the Gnome a semi-rotary engine: it is a very light engine, but could not something still lighter and simpler be devised?

THE SCREW PROPELLER

The screw propeller, equally with the engine, is a factor of the first importance. A great deal has been written about the screw, and multitudinous calculations have been made by mathematicians, but I do not think that much fresh light has been thrown on the subject since the investigations of

* Later Sir Dugald Clerk.

William Froude. The fact is that the motion of the water in the neighbourhood of the screw is extremely complex and defies exact calculation.

What Froude established experimentally was that a well-made screw of about 0·3 ratio of projected blade to disc area reached about 70 per cent. of efficiency. He also established the laws of resistance of planes of different lengths, and with various degrees of roughness of surface, when dragged edge-ways in water, which laws are of great importance in propeller design.

There are four chief factors of loss in propellers: First of all, loss by slip; secondly, loss by skin friction of the blades on the water; thirdly, a loss from centrifugal force on the thin film of water carried round by the surface of the blade urging it outwards; and fourthly, a loss from cavitation, which should only occur in fast ships provided the screw and stern lines of the vessel are reasonably well designed.

The loss by slip and the loss by skin friction are, in design, generally made similar in quantity, because if the former is reduced by increasing the diameter of the propeller, the latter is increased; therefore the laws of maxima and minima demand some approximation to equality. The loss by centrifugal force is probably very small, and may be neglected in all cases, for circumferential corrugations on the blades have been tried to catch or impede it, and these have shown that the extra skin friction of the corrugations more than neutralises the gain, so that in slow ships at all events we need only consider slip and skin friction as established experimentally by Froude.

Since Froude's work was completed, the power and speed of ships has enormously increased. Thornycroft and Barnaby were the first to observe and recognise that cavitation had to be reckoned with. As judged by close observation of screws in our experimental tanks, it appears that the most active cause producing cavitation is the pressure of the blades on the water to drive the vessel, which causes a transverse flow over the leading edge and tips from front to back, and also a diminished pressure over the whole of the backs resulting in cavities over the backs and vacuous vortex spaces or spirals emanating from the cavities. These are often very permanent and survive for several diameters abaft the propeller.

The other cause is due to the thickness of the blades and the inability of the water at high surface speeds to follow around the curvature of the backs.

The panacea adopted has been to widen the blades to from 0·5 to 0·7 surface ratio, and to make them as thin as considerations of strength will allow. The widening of the blades obviously entails more loss by skin friction, but, following from Froude's laws, it is in a less proportion than such widening; also the loss is less in large screws than in small, and for these reasons some sacrifice in propeller efficiency is always incurred in fast vessels.

The advocates of gearing have justly made claim on this score, though in my opinion the electrical gear advocates have overstated the probable gain. It is, however, obvious, since skin friction is the cause of the increased loss, that by the use of gearing a lower rate of revolution may be chosen without increase of engine weights; consequently with the same diameter and size of screw and a coarser pitch, the surface velocity will be less and the skin friction reduced nearly as the square of such reduction. Very few data, however, exist on the subject, because in fast ships the engine (whether reciprocating or turbine) has to run as fast as possible because weight is of primary importance in such vessels, but with gearing interposed the only increase of weight is in the line shafting and propellers, and is comparatively small.

ROLLING OF SHIPS

Apparatus for reducing the rolling of ships has attracted more attention of late, following on the work of Mr Frahm, and is now applied to some large mercantile and war vessels. The names of William Froude, White, Thornycroft, Tower, Bertin, Watts, Biles and others, are associated with the early attack on this problem which commenced about the year 1882. Devices for quelling oscillations by chambers containing viscous liquid had been well known previously. A familiar experiment is the contrast between the behaviour of a raw and a hard boiled egg when placed on the table. Sir William Thomson* had also adopted similar means to steady his compasses.

Watts, in his paper to the Institution of Naval Architects in 1883 on "A Method of Reducing the Rolling of Ships at Sea," described the athwart ships tanks fitted in the *Inflexible*, their theory and behaviour. Thornycroft, ten years later, fitted a segmental pivoted weight moved by a hydraulic cylinder actuated by a valve admitting water under pressure to the cylinder, the valve itself being controlled by a long period pendulum. With both apparatus the resulting force introduced to check the rolling is similar, the water in the chamber in the one case, and the moving weight in the other, find themselves always on the lifting and high side of the vessel, and so reduce or annul the rolling. These devices were partially successful, but did not come into general use.

Schlick, about the year 1903, attacked the problem of steadying the vessel by the gyroscope, as Tower had successfully steadied a gun platform many years before. Both men utilised the well-known resistance of the gyroscope to changes of its plane of rotation. Schlick used a large gyroscope of steel rotated at a high speed by an electric motor and hung in athwartship trunnions, the movement of the gyroscopic axis being controlled by centralising springs and retarded by a hydraulic dashpot. Schlick achieved successful results in several vessels, but the system has scarcely, as yet, proceeded beyond the experimental stage.

* Lord Kelvin.

Schlick's method may be described as the use of retarded precession for the steadying of stable things such as ships, just as Brennan uses artificially accelerated precession for the maintenance of unstable things like locomotives in an upright position on one rail.

Four years ago Dr Hermann Frahm, of Messrs Blohm and Voss, of Hamburg, took up the subject of anti-rolling tanks in vessels on similar lines to those described by Watts. He was impressed with the importance of the beautiful and well-known properties of resonance between vibrating systems (by the way, properties utilised and essential not only in musical instruments but also in wireless telegraphy). The principle of resonance when applied to the steadying of vessels demands that the period of oscillation of the water in the tanks must approximate to or be a little quicker than the period of metacentric roll of the vessel and the motion of the water, must be suitably damped, to secure the best result.

CONCLUSION

In conclusion, I should like to address a few remarks to our junior members. Things have moved ahead in the last twenty-eight years; competition for success in our profession grows keener year by year; the field of work widens and grows more complex, demanding not only more concentration of thought but also a better knowledge of the work of other men, and of the fields they have explored, and of their store of knowledge.

The old rule of thumb methods have been discarded; the discoveries and data made and tabulated by physicists, chemists, and metallurgists are eagerly sought by the engineer, and, as far as possible, utilised by him in his methods and designs.

The man of great energy and mental powers, the man of genius, will undoubtedly struggle to the top in spite of inadequate education and opportunities, but the average man is very largely dependent on his training, and without it has a very much worse chance of success; and even the man of poor ability may do well if he prepares himself with due diligence.

In my experience I have come to the conclusion that the old adage that "pride goeth before a fall" describes in this day the commonest cause of failure in young men. One form it takes, and this I mention as a warning, is that, fresh from technical school or college, he often thinks he knows practically everything and more than quiet men who have given the better part of their lives to the work in hand; he does not realise that many of the facts or things stated shortly in text books may have taken a lifetime to discover. A few years, however, of practical and responsible work generally brings about a juster appreciation of the work of others, and a better proportioned estimate of his own individual merit in relation to others; he also finds that the acquisition of knowledge does not end with the school or college, but must be continued through life.

PRESIDENTIAL ADDRESS

British Association, Bournemouth, 1919

We are gathered together at a time when, after a great upheaval, the elemental conditions of organisation of the world are still in flux, and we have to consider how to mould and influence the recrystallisation of these elements into the best forms and most economic rearrangements for the benefit of civilisation. That the British Association has exerted a great influence in guiding the nation towards advancement in the Sciences and Arts in the most general sense there can be no question, and of this we may be assured by a study of its proceedings in conjunction with the history of contemporary progress. Although the British Association cannot claim any paramount prerogative in this good work, yet it can certainly claim to provide a free arena for discussion where in the past new theories in Science, new propositions for beneficial change, new suggestions for casting aside fetters to advancement in Science, Art and Economics have first seen the light of publication and discussion.

For more than half a century it has pleaded strongly for the advancement of Science and its application to the Arts. In the yearly volume for 1855 will be found a report in which it is stated that "The Objects for which the Association was established have been carried out in three ways: First, by requisitioning and printing reports on the present state of different branches of Science; secondly, by granting sums of money to small committees or individuals, to enable them to carry on new researches; thirdly, by recommending the Government to undertake expeditions of discovery, or to make grants of money for certain and national purposes, which were beyond the means of the Association". As a matter of fact it has, since its commencement, paid out of its own funds upwards of £80,000 in grants in aid of research.

Developments prior to the war

It is twenty-nine years since an engineer, Sir Frederick Bramwell, occupied this chair and discoursed so charmingly on the great importance of the next to nothing, the importance of looking after little things which, in engineering, as in other walks of life, are often too lightly considered.

The advances in engineering during the last twenty years are too many and complex to allow of their description, however short, being included in one Address, and, following the example of some of my predecessors in this chair, I shall refer only to some of the most important features of this wide subject. I feel that I cannot do better than begin by quoting from a speech made recently by Lord Inchcape, when speaking on the question of the nationalisation of coal: "It is no exaggeration to say that coal has been the maker of modern Britain, and that those who discovered and developed the methods of working it have done more to determine the bent of British activities and the form of British society than all the Parliaments of the past hundred and twenty years".

James Watt

No excuse is necessary for entering upon this theme, because this year marks the hundredth anniversary of the death of James Watt, and in reviewing the past, it appears that England has gained her present proud position by her early enterprise and by the success of the Watt steam engine, which enabled her to become the first country to develop her resources in coal, and led to the establishment of her great manufactures and her immense mercantile marine.

The laws of steam which James Watt discovered are simply these: That the latent heat is nearly constant for different pressures within the ranges used in steam engines, and that, consequently, the greater the steam pressure and the greater the range of expansion the greater will be the work obtained from a given amount of steam. Secondly, as may now seem to us obvious, that steam from its expansive force will rush into a vacuum. Having regard to the state of knowledge at the time, his conclusions appear to have been the result of close and patient reasoning by a mind endowed with extraordinary powers of insight into physical questions, and with the faculty of drawing sound practical conclusions from numerous experiments devised to throw light on the subject under investigation. His resource, courage and devotion were extraordinary.

In commencing his investigations on the steam engine he soon discovered that there was a tremendous loss in the Newcomen engine, which he thought might be remedied. This was the loss caused by condensation of the steam on the cold metal walls of the cylinder. He first commenced by lining the walls with wood, a material of low thermal conductivity. Though this improved matters, he was not satisfied; his intuition probably told him that there should be some better solution of the problem, and doubtless he made many experiments before he realised that the true solution lay in a condenser separate from the cylinder of the engine. It is easy after discovery to say, "How obvious and how simple", but many of us here know how difficult is any step of advance when shrouded by unknown surroundings, and we can well appreciate the courage and the amount of investigation necessary before James Watt thought himself justified in trying the separate condenser. But to us now, and to the youngest student who knows the laws of steam as formulated by Carnot, Joule, and Kelvin, the separate condenser is the obvious means of constructing an economical condensing engine.

Watt's experiments led him to a clear view of the great importance of securing as much expansion as possible in his engines. The materials and appliances for boiler and machine construction were at that time so undeveloped that steam pressures were practically limited to a few pounds above atmospheric pressure. The cylinders and pistons of his engines were not constructed with the facility and accuracy to which we are now accustomed, and chiefly for these reasons expansion ratios of from two- to three-fold were the usual practice. Watt had given to the world an engine which consumed from five to seven pounds of coal per horse-power hour, or one-quarter of the fuel previously used by any engine. With this consumption of fuel its field under the conditions prevailing at the time was practically unlimited. What need was there, therefore, for commercial reasons, to endeavour still further to improve the engine at the risk of encountering fresh difficulties and greater commercial embarrassments? The course was rather for him and his partners to devote all their energy to extending the adoption of the engine as it stood, and this they did, and to the Watt engine, consuming from five to seven pounds of coal per horse-power hour, mankind owes the greatest permanent advances in material welfare recorded in history.

With secondary modifications, it was the prime mover in most general use for eighty years—i.e. till the middle of last century. It remained for others to carry the expansion of steam still further in the compound, triple, and, lastly, in the quadruple expansion engine, which is the most economical reciprocating engine of to-day.

Watt had considered the practicability of the turbine. He writes to his partner, Boulton, in 1784: "The whole success of the machine depends on the possibility of prodigious velocities. In short, without God makes it possible for things to move them one thousand feet per second, it cannot do us much harm". The advance in tools of precision, and a clearer knowledge of the dynamics of rotating bodies, have now made the speeds mentioned by Watt feasible, and indeed common, everyday practice.

Turbines

The turbine of to-day carries the expansion of steam much further than has been found possible in any reciprocating engine, and owing to this property it has surpassed it in economy of coal, and it realises to the fullest extent Watt's ideal of the expansion of steam from the boiler to the lowest vapour pressure obtainable in the condenser.

Among the minor improvements which in recent years have conduced to a higher efficiency in turbines are the more accurate curvature of the blades to avoid eddy losses in the steam, the raising of the peripheral velocities of the blades to nearly the velocity of the steam impinging upon them, and details of construction to reduce leakages to a minimum. In turbines of 20,000 to 30,000 horse-power 82 per cent. of the available energy in the steam is now obtainable as brake horse-power; and with a boiler efficiency of 85 per cent. the thermo-dynamic efficiency from the fuel to the electrical output of the alternator has reached 23 per cent., and shortly may reach 28 per cent., a result rivalling the efficiency of internal combustion engines worked by producer gas.

During the twenty years immediately preceding the war turbo generators had increased in size from 500 kilowatts to 25,000 kilowatts, and the consumption of steam had fallen from 17 lb. per kilowatt hour to 10·3 lb. per kilowatt hour. Turbines have become the recognised means of generating electricity from steam on a large scale, although they have not superseded the Watt engine for pumping mines or the drawing of coal, except as a means for generating electricity for these purposes. In the same period the engine power in the mercantile marine had risen from 3900 of the *King Edward* to 75,000 of the *Mauretania*.

As regards the Royal Navy, the engine power of battleships, prior to the war, had increased from 12,000 indicated horse-power to 30,000 shaft horse-power, while the speed advanced from 17 knots to 23 knots, and during the war, in ships of the *Queen Elizabeth* class, the power amounted to 75,000 shaft horse-power, with a speed of 25 knots. In cruisers similar advances were made. The indicated horse-power of the *Powerful* was 25,000, while the shaft horse-power of the *Queen Mary* was 78,000, with a speed of 28 knots. During the war the power obtained with geared turbines in the *Courageous* class was 100,000 shaft horse-power with a speed of 32 knots, the maximum power transmitted through one gear wheel being 25,000 horse-power, and through one pinion 15,500 horse-power, while in destroyers, speeds up to 39 knots have been obtained. The aggregate horse-power of war and mercantile turbined vessels throughout the world is now about 35 millions.

These advances in power and speed have been made possible mainly by the successive increase in economy and diminution of weight derived from the replacement of reciprocating engines by turbines direct coupled to the propellers, and, later, by the introduction of reduction gearing between the turbines and the propellers; also by the adoption of water-tube boilers and of oil fuel. With these advances the names of Lord Fisher, Sir William White, and Sir Henry Oram will always be associated.

The British Navy has led the world for a century and more. Lord Fisher has recently said that many of the ships are already obsolete and must soon be replaced if supremacy is to be maintained; and there can be no question that to guide the advance and development on the best lines, continuous scientific experiment, though costly at the time, will prove the cheapest in the long run.

The Work of Sir William White

With the great work of the Royal Navy fresh in our minds, we cannot but recall the prominent part taken by the late Sir William White in its construction. His sudden death, when President-elect for 1913, lost to the nation and to the Association the services of a great naval architect who possessed remarkable powers of prevision and dialectic. He was Chief Constructor to the Admiralty from 1885 to 1901, and largely to him was due the efficiency of our vessels in the Great War.

White often referred to the work of Brunel as the designer of the *Great Eastern*, and spoke of him as the originator of the cellular construction of the bottoms of ships, since universally adopted, as a means of strengthening the hull and for obtaining additional safety in case of damage. Scott-Russell was the builder of this great pioneer vessel, the forerunner of the Atlantic liners, and the British Association may rightly feel satisfaction in having aided him when a young man by pecuniary grants to develop his researches into the design and construction of ships and the wave-line form of hull which he originated, a form of special importance in paddle-wheel vessels.

So much discussion has taken place in the last four years as to the best construction of ship to resist torpedo attacks that it is interesting to recall briefly at the present time what was said by White in his Cantor Lectures to the Royal Society of Arts in 1906: "Great attention has been bestowed upon means of defence against underwater torpedo attacks. From the first introduction of torpedoes it was recognised that extreme watertight sub-division in the interior of warships would be the most important means of defence. Experiments have been made with triple watertight skins forming double cellular sides, the compartments nearest the outer bottom being filled, in some cases, with water, coal, cellulose, or other materials. Armour plating has been used both on the outer bottom and on inner skins". He also alluded to several Russian ships which were torpedoed by the Japanese, and he concluded by saying: "Up to date the balance of opinion has favoured minute watertight subdivisions and comparatively thin water-tight compartments, rather than the use of internal armour, whose use, of course, involves large expenditure of weight and cost".

The present war has most amply confirmed his views and conclusions, then so lucidly and concisely expressed.

While on the subject of steamships, it may perhaps be opportune to say one word as to their further development. The size of ships steadily increasing up to the time of the war, resulting in a reduction of power required to propel them per ton of displacement. On the other hand, thanks to their greater size and more economical machinery, speeds have been increased when the traffic has justified the greater cost. The limiting factor to further increase in size is the depth of water in the harbours. With this restriction removed there is no obstacle to building ships up to 1000 feet in length or more, provided the volume and character of the traffic are such as to justify the capital outlay.

Tungsten Steel

Among other important pre-war developments that have had a direct bearing upon the war, mention should be made of the discovery and extensive use of alloys of steel. The wonderful properties conferred upon steel by the addition of tungsten were discovered by Mushet* in 1868, and later this alloy was investigated and improved by Maunsel White and Taylor, of Philadelphia. The latter showed that the addition of tungsten to steel has the following effect: That after the steel has been quenched at a very high temperature near its melting point it can be raised to a much higher temperature than is possible with ordinary carbon tool steel, without losing its hardness and power of cutting metal. In other words, it holds the carbon more tenaciously in the hardened state, and hence tungsten steel tools, even when red hot, can cut ordinary mild steel. It has revolutionised the design of machine tools and has increased the output on heavy munition work by 100 per cent., and in ordinary engineering by 50 per cent.

The alloys of steel and manganese with which the name of Sir Robert Hadfield is associated have proved of utility in immensely increasing the durability of railway and tramway points and crossings, and for the hard teeth of machinery for the crushing of stone and other materials, and, in fact, for any purposes where great hardness and strength are essential.

Investigation of Gaseous Explosions

Brief reference must also be made—and it will be gratifying to do so— to the important work of one of the Committees of the British Association appointed in 1908, under the chairmanship of the late Sir William Preece, for the investigation of gaseous explosions, with special reference to temperature. The investigations of the Committee are contained in seven yearly reports up to 1914. Of the very important work of the Committee I wish to refer to one investigation in particular, which has proved to be a guiding star to the designers and manufacturers of internal combustion engines in this country. The members of the Committee more directly

* Who has not been sufficiently credited with his share in making the Bessemer process a practical success.

associated with this particular investigation were Sir Dugald Clerk,*
Professor Callendar,† and the late Professor Bertram Hopkinson.

The investigation showed that the intensity of the heat radiated by the
incandescent gases to the walls of the cylinder of a gas engine increases
with the size of the cylinder, the actual rate of this increase being approximately proportional to the square root of the depth of the radiating incandescent gas; the intensity was also shown to increase rapidly with the
richness of the gas. It suffices now to say that the heat in a large cylinder
with a rich explosive mixture is so intense that the metal eventually cracks.
The investigation shows why this occurs, and by doing so has saved
enormous sums to the makers of gas and oil engines in this country, and
has led them to avoid the large cylinder, so common in Germany before
the war, in favour of a multiplicity of smaller cylinders.

SCIENCE AND THE WAR

In coming to this section of my Address I am reminded that in the course
of his Presidential Address to Section G, in 1858, Lord Rosse said: "Another object of the Mechanical Section of the Association has been effected
—the importance of engineering science in the service of the State has been
brought more prominently forward. There seems, however, something still
wanting. Science may yet do more for the Navy and Army if more called
upon".

Comparatively recently, too, Lord French remarked: "We have failed
during the past to read accurately the lessons as regards the fighting of the
future which modern science and invention should have taught us".

In view of the eminent services which scientists have rendered during
the war, I think that we may be justified in regarding the requirement
stated by Lord Rosse as having at last been satisfied, and also in believing
that such a criticism as Lord French rightly uttered will not be levelled
against the country in the future.

Though British men of Science had not formerly been adequately recognised in relation to war and the safety of their country, yet at the call
of the sailors and the soldiers they whole-heartedly, and with intense zeal,
devoted themselves to repair the negligence of the past, and to apply their
unrivalled powers and skill to encounter and overcome the long-standing
machinations of the enemy. They worked in close collaboration with the
men of Science of the Allied Nations, and eventually produced better war
material, chemicals, and apparatus of all kinds for vanquishing the enemy
and the saving of our own men than had been devised by the enemy during
many many years of preparation planned on the basis of a total disregard
of treaties and the conventions of war.

* Obit: Nov. 1932.   † Obit: Jan. 1930.

Four years is too short a time for much scientific invention to blossom to useful maturity, even under the forced exigencies of war and Government control. It must be remembered that in the past the great majority of new discoveries and inventions of merit have taken many years—sometimes generations—to bring them into general use. It must also be mentioned that in some instances discoveries and inventions are attributable to the general advance in Science and the Arts which has brought within the region of practical politics an attack on some particular problem. So the work of the scientists during the war has perforce been directed more to the application of known principles, trade knowledge, and properties of matter to the waging of war, than to the making of new and laborious discoveries; though, in effecting such applications, inventions of a high order have been achieved, some of which promise to be of great usefulness in time of peace.

The advance of Science and the Arts in the last century had, however, wrought a great change in the implements of war. The steam engine, the internal combustion engine, electricity, and the advances in metallurgy and chemistry had led to the building up of immense industries which, when diverted from their normal uses, produced unprecedented quantities of war material for the enormous armies, and also for the greatest Navy which the world has ever seen.

The destructive energy in the field and afloat has multiplied many hundredfold since the time of the Napoleonic wars; both before and during the war the size of guns and the efficiency of explosives and shell increased immensely, and many new implements of destruction were added. Modern Science and Engineering enabled armies unprecedented in size, efficiency and equipment to be drawn from all parts of the world and to be concentrated rapidly in the fighting line.

To build up the stupendous fighting organisation, ships have been taken from their normal trade routes, locomotives and material from the home railways, the normal manufactures of the country have been largely diverted to munitions of war; the home railways, tramways, roads, buildings and constructions, and material of all kinds have been allowed to depreciate. The amount of depreciation in roads and railways alone has been estimated at 400 millions per annum at present prices. Upon the community at home a very great and abnormal strain has been thrown, notwithstanding the increased output per head of the workers derived from modern methods and improved machinery. In short, we have seen for the first time in history nearly the whole populations of the principal contending nations enlisted in intense personal and collective effort in the contest, resulting in unprecedented loss of life and destruction of capital.

A few figures will assist us to realise the great difference between this war and all preceding wars. At Waterloo, in 1815, 9044 artillery rounds were fired, having a total weight of 37·3 tons, while on one day during the last offensive in France, on the British Front alone, 943,837 artillery rounds were fired, weighing 18,080 tons—over 100 times the number of rounds, and 485 times the weight of projectiles. Again, in the whole of the South African War, 273,000 artillery rounds were fired, weighing approximately 2800 tons; while during the whole war in France, on the British Front alone, over 170 million artillery rounds were fired, weighing nearly $3\frac{1}{2}$ million tons—622 times the number of rounds, and about 1250 times the weight of projectiles.

However great these figures in connection with modern land artillery may be, they become almost insignificant when compared with those in respect of a modern naval battle squadron. The *Queen Elizabeth* when firing all her guns discharges 18 tons of metal and develops 1,870,000 foot-tons of energy. She is capable of repeating this discharge once every minute, and when doing so develops by her guns an average of 127,000 effective horse-power, or more than one-and-a-half times the power of her propelling machinery. This energy is five times greater than the maximum average energy developed on the Western Front by British guns. Furthermore, if all her guns were fired simultaneously, they would for the instant be developing energy at the rate of 13,132,000 horse-power. From these figures we can form some conception of the vast destructive energy developed in a modern naval battle.

ENGINEERING AND THE WAR

Sound-ranging and Listening Devices

Probably the most interesting development during the war has been the extensive application of sound-listening devices for detecting and localising the enemy. The Indian hunter puts his ear to the ground to listen for the sound of the footsteps of his enemy. So in modern warfare science has placed in the hands of the sailor and soldier elaborate instruments to aid the ear in the detection of noises transmitted through earth, water, air, or ether, and also in some cases to record these sounds graphically or photographically, so that their character and the time of their occurrence may be tabulated.

The sound-ranging apparatus by which the position of an enemy gun can be determined from electrically recorded times at which the sound wave from the gun passes over a number of receiving stations, has enabled our artillery to concentrate their fire on the enemy's guns, and often to destroy them.

The French began experimenting in September 1914 with methods of locating enemy guns by sound. The English section began work in October

1915, adopting the French methods in the first instance. By the end of 1916 the whole Front was covered, and sound-ranging began to play an important part in the location of enemy batteries. During 1917 locations by sound-ranging reached about 30,000 for the whole army, this number being greater than that given by any other means of location. A single good set of observations could be relied upon to give the position of an enemy gun to about 50 yards at 7000 yards range. It could also be carried on during considerable artillery activity.

The apparatus for localising noises transmitted through the ground has been much used for the detection of enemy mining and counter-mining operations. Acoustic tubes, microphones, and amplifying valves have been employed to increase the volume of very faint noises.

For many years before the war the Bell Submarine Signalling Company, of which Sir William White was one of the early directors, used submerged microphones for detecting sound transmitted through the water, and a submerged bell for sending signals to distances up to one mile. With this apparatus passing ships could be heard at a distance of nearly a mile when the sea was calm and the listening vessel stationary.

Of all the physical disturbances emitted or produced by a moving submarine, those most easily detected, and at the greatest distance, are the pressure waves set up in the water by vibrations produced by the vessel and her machinery. A great variety of instruments have been devised during the war for detecting these noises, depending on microphones and magnetophones of exceedingly high sensitivity. Among them may be particularly mentioned the hydrophones devised by Captain Ryan and Professor Bragg*, being adaptations of the telephone transmitter to work in water, instead of air. These instruments, when mounted so as to rotate, are directional, being insensitive to sound waves whose front is perpendicular to the plane of the diaphragm, and giving the loudest sound when the diaphragm is parallel to the wave front.

Another preferable method for determining direction is to use two hydrophones coupled to two receivers, one held to each ear. This is called the biaural method, and enables the listener to recognise the direction from which the sound emanates.

When the vessel is in motion or the sea is rough the water noises from the dragging of the instrument through the water and from the waves striking the ship drown the noises from the enemy vessel, and under such conditions the instruments are useless. The assistance of eminent biologists was of invaluable help at this juncture. Experiments were made with sealions by Sir Richard Paget, who found that they have directional hearing under water up to speeds of 6 knots. Also Professor Keith† explained the

_____
* Now Sir William Bragg, K.B.E., F.R.S.   † Now Sir Arthur Keith, Kt., F.R.S.

construction of the hearing organs of the whale, the ear proper being capillary tube, too small to be capable of performing any useful function in transmitting sound to the relatively large aural organs, which are deep set in the head. The whale therefore hears by means of the sound waves transmitted through the substance of the head. It was further seen that the organs of hearing of the whale to some degree resembled the hydrophone.

The course now became clear. Hollow towing bodies in the form of fish or porpoises were made of celluloid, varnished canvas, or very thin metal, and the hydrophone suitably fixed in the centre of the head. The body is filled with water, and the cable towing the fish contains the insulated leads to the observer on board the vessel. When towed at some distance behind the chasing ship disturbing noises are small, and enemy noises can be heard up to speeds of 14 knots, and at considerable distances. Thermionic amplifying valves have been extensively used, and have added much to the sensitiveness of the hydrophone in its many forms.

After the loss of the *Titanic* by collision with an iceberg, Lewis Richardson was granted two patents in 1912 for the detection of above-water objects by their echo in the air, and under-water objects by their echo transmitted through the water. The principles governing the production and the concentration of beams of sound are described in his specifications, and he recommends frequencies ranging from 4786 to 100,000 complete vibrations per second, and also suggests that the rate of approach or recession from the object may be determined from the difference in the pitch of the echo from the pitch of the blast sent out. Hiram Maxim also suggested similar apparatus a little later.

The echo method of detection was not, however, practically developed until French and English scientists, with whom was associated Professor Langevin, of the College de France, realising its importance for submarine detection, brought the apparatus to a high degree of perfection and utility shortly before the Armistice. Now, with beams of high-frequency sound waves, it is possible to sweep the seas for the detection of any submerged object, such as icebergs, submarines, surface vessels, and rocks; they may also be used to make soundings. The apparatus enables a chasing ship to pick up and close in on a submarine situated more than a mile away.

The successful development of sound-ranging apparatus on land led to the suggestion by Professor Bragg that a modified form could be used to locate under-water explosions. It has been found that the shock of an explosion can be detected hundreds of miles from its source by means of a submerged hydrophone, and that the time of the arrival of the sound wave can be recorded with great precision. At the end of the war the sound-ranging stations were being used for the detection of positions at sea, required for strategical purposes. The same stations are now being used extensively for the determination of such positions at sea as light-vessels, buoys which indicate channels, and obstructions such as sunken ships. By this means ships steaming in fog can be given their positions with accuracy for ranges up to 500 miles.

Among the many other important technical systems and devices brought out during the war which will find useful application under peace conditions as aids to navigation I may mention directional wireless, by which ships and aircraft can be given their positions and directed.

Leader gear, first used by the Germans to direct their ships through their minefields, and subsequently used by the Allies, consists of an insulated cable laid on the bottom of the sea, earthed at the further end, through which an alternating current is passed. By means of delicate devices installed on a ship, she is able to follow the cable at any speed with as much precision as a railless electric 'bus can follow its trolley wire. Cables up to 50 miles long have been used, and this device promises to be invaluable to ships navigating narrow and tortuous channels and entering or leaving harbours in a fog.

Aircraft

It may be justly said that the development in aircraft design and manufacture is one of the astonishing engineering feats of the war. In August 1914 the British Air Services possessed a total of 272 machines, whereas in October 1918, just prior to the Armistice, the Royal Air Force possessed over 22,000 effective machines. During the first twelve months of the war the average monthly delivery of aeroplanes to our Flying Service was fifty, while during the last twelve months of the war the average deliveries were 2700 per month. So far as aero-engines are concerned, our position in 1914 was by no means satisfactory. We depended for a large proportion of our supplies on other countries. In the Aerial Derby of 1913, of the eleven machines that started, not one had a British engine. By the end of the war, however, British aero-engines had gained the foremost place in design and manufacture, and were well up to requirements as regards supply. The total horse-power produced in the last twelve months of the war approximated to eight millions of brake horse-power, a figure quite comparable with the total horse-power of the marine engine output of the country *.

Much might be written on the progress in aircraft, but the subject will be treated at length in the sectional papers. In view of the recent trans-Atlantic flights, however, I feel that it may be opportune to make the following observations on the comparative utility of aeroplanes and airships for commercial purposes. In the case of the aeroplane, the weight

* See Lord Weir's Paper read at the Victory Meeting of the North-East Coast Institution of Engineers and Shipbuilders, July 1919.

per horse-power increases with the size, other things being equal. This increase, however, is met to some extent by a multiplicity of engines, though in the fusilage the increase remains.

On the other hand, with the airship the advantage increases with the size, as in all ships. The tractive effort per ton of displacement diminishes in inverse proportion to the dimensions, other things, including the speed, being the same. Thus, an airship of 750 feet length and 60 tons displacement may require a tractive force of 5 per cent., or 3 tons, at 60 miles per hour; while one of 1500 feet length and $8 \times 60 = 480$ tons displacement would only require $2\frac{1}{2}$ per cent. $\times 480 = 12$ tons at the same speed, and would carry fuel for double the distance.

With the same proportion of weight of hull to displacement, the larger airship would stand double the wind pressure, and would weather storms of greater violence and hailstones of greater size. It would be more durable, the proportional upkeep would be lower, and the proportional loss of gas considerably less. In other words, it would lose a less proportion of its buoyancy per day. It is a development in which success depends upon the project being well thought out and the job being thoroughly well done. The equipment of the airsheds with numerous electric haulage winches, and all other appliances to make egress and ingress to the sheds safe from danger and accident, must be ample and efficient.

The airship appears to have a great future for special commerce where time is a dominant factor and the demand is sufficient to justify a large airship. It has also a great field in the opening up of new countries where other means of communication are difficult. The only limitation to size will be the cost of the airship and its sheds, just as in steam vessels it is the cost of the vessels and the cost of deepening the harbours that limit the size of Atlantic liners.

Such developments generally take place slowly, otherwise failures occur —as in the case of the *Great Eastern*—and it may be many years before the airship is increased from the present maximum of 750 to 1500 feet with success, but it will assuredly come. If, however, the development is subsidised or assisted by Government, incidental failures may be faced with equanimity and very rapid development accomplished. In peace time the seaplane, aeroplane, and airship will most certainly have their uses. But, except for special services of high utility, it is questionable whether they will play more than a minor part as compared with the steamship, railway, and motor transport.

Electricity

The supply and use of electricity has developed rapidly in recent years. For lighting it is the rival of gas, though each has its advantages. As a means of transmitting power over long distances it has no rival, and its efficiency is so high that when generated on a large scale and distributed over large areas it is a cheap and reliable source of power for working factories, tramways, suburban railways, and innumerable other purposes, including metallurgical and chemical processes. It is rapidly superseding locally generated steam-power, and is a rival to the small and moderate-sized gas and oil engine. It has made practicable the use of water-power through the generation of electricity in bulk at the natural falls, from which the power is transmitted to the consumers, sometimes at great distances.

Fifteen years ago electricity was generated chiefly by large reciprocating steam engines, direct coupled to dynamos or alternators, but of late years steam turbines have in most instances replaced them, and are now exclusively used in large generating stations, because of their smaller cost and greater economy in fuel. The size of the turbines may vary from a few thousand horse-power up to about 50,000 horse-power.* At the end of last year the central electric stations in the United Kingdom contained plant aggregating $2\frac{3}{4}$ million kilowatts, 79 per cent. of which was driven by steam turbines.

Much discussion has taken place as to the most economical size of generating stations, their number, the size of the generating units, and the size of the area to be supplied. On the one hand, a comparatively small number of very large or super-stations, instead of a large number of moderate-sized stations dotted over the area, results in a small decrease in the cost of production of the electricity, because in the super-stations larger and slightly more economical engines are employed, while the larger stations permit of higher organisation and more elaborate labour-saving appliances. Further, if in the future the recovery of the by-products of coal should become a practical realisation as part of the process in the manufacture of the electric current, the larger super-stations present greater facilities than the smaller stations. On the other hand, super-stations involve the transmission of the electricity over greater distances, and consequently greater capital expenditure and cost of maintenance of mains and transmission apparatus, and greater electrical transmission losses, while the larger generating unit takes longer to overhaul or repair, and consequently a larger percentage of spare plant is necessary.

The greatest element in reducing the cost of electricity is the provision of a good load factor; in other words, the utilisation of the generating plant and mains to the greatest extent during the twenty-four hours of each day throughout the year. This is a far more important consideration than the size of the station, and it is secured to the best advantage in most cases by a widespread network of mains, supplying a diversity of consumers and

* [1934. Chicago 150,000 kW. Ed.]

uses, each requiring current at different times of the day. The total load of each station being thus an average of the individual loads of a number of consumers is, in general, far less fluctuating than in the case of small generating and distributing systems, which supply principally one class of consumer, a state of affairs that exists in London, for instance, at the present time. It is true that there may be exceptional cases, such as at Kilmarnock, where a good load factor may be found in a small area, but in this case the consumers are chiefly mills, which require current for many hours daily.

There is no golden rule to secure cheap electricity. The most favourable size, locality, and number of generating stations in each area can only be arrived at by a close study of the local conditions, but there is no doubt that, generally speaking, to secure cheap electricity a widespread network of mains is in most cases a very important, if not an essential, factor.

The electrification of tramways and suburban railways has been an undoubted success where the volume of traffic has justified a frequent service, and it has been remarkable that where suburban lines have been worked by frequent and fast electrical trains there has resulted a great growth of passenger traffic. The electrification of main line railways would no doubt result in a saving of coal; at the same time, the economical success would largely depend on the broader question as to whether the volume of the traffic would suffice to pay the working expenses, and provide a satisfactory return on the capital.

Municipal and company generating stations have been nearly doubled in capacity during the war to meet the demand from munition works, steel works, chemical works, and for many other purposes. The provision of this increased supply was an enormous help in the production of adequate munitions. At the commencement of the war there were few steel electric furnaces in the country; at the end of last year 117 were at work, producing 20,000 tons of steel per month, consisting chiefly of high-grade ferro alloys used in munitions.

THE FUTURE

The nations who have exerted the most influence in the war have been those who have developed to the greatest extent their resources, their manufactures, and their commerce. As in the war, so in the civilisation of mankind. But, viewing the present trend of developments in harnessing water-power and using up the fuel resources of the world for the use and convenience of man, one cannot but realise that, failing new and unexpected discoveries in science, such as the harnessing of the latent molecular and atomic energy in matter, as foreshadowed by Clerk Maxwell, Kelvin, Rutherford, and others, the great position of England cannot be maintained for an indefinite period. At some time more or less remote—long before the exhaustion of our coal—the population will gradually migrate to those countries where the natural sources of energy are the most abundant.

Water-power and Coal

The amount of available water-power in the British Isles is very small as compared with the total in other countries. According to the latest estimates, the total in the British Isles is under $1\frac{1}{2}$ million horse-power, whereas Canada alone possesses over 20 millions, of which over 2 millions have already been harnessed. In the rest of the British Empire there are upwards of 30 millions and in the remainder of the world at least 150 millions, so that England herself possesses less than 1 per cent. of the water-power of the world. Further, it has been estimated that she only possesses $2\frac{1}{2}$ per cent. of the whole coal of the world. To this question I would wish to direct our attention for a few minutes.

I have said that England owes her modern greatness to the early development of her coal. Upon it she must continue to depend almost exclusively for her heat and source of power, including that required for propelling her vast mercantile marine. Nevertheless, she is using up her resources in coal much more rapidly than most other countries are consuming theirs, and long before any near approach to exhaustion is reached her richer seams will have become impoverished, and the cost of mining so much increased that, given cheap transport, it might pay her better to import coal from richer fields of almost limitless extent belonging to foreign countries, and workable at a much lower cost than her own.

Let us endeavour to arrive at some approximate estimate of the economic value of the principal sources of power. The present average value of the royalties on coal in England is about $6d$. per ton, but to this must be added the profit derived from mining operations after paying royalties and providing for interest on the capital expended and for its redemption as wasting capital. After consultation with several leading experts in these matters, I have come to the conclusion that about $1s$. per ton represents the pre-war market value of coal in the seams in England.

It must, however, be remembered that, in addition, coal has a considerable value as a national asset, for on it depends the prosperity of the great industrial interests of the country, which contribute a large portion of the wealth and revenue. From this point of view the present value of unmined coal seems not to have been sufficiently appreciated in the past, and that in the future it should be better appraised at its true value to the nation.

This question may be viewed from another aspect by making a comparison of the cost of producing a given amount of electrical power from coal and from water-power. Assuming that 1 horse-power of electrical energy maintained for one year had a pre-war value of £5, and that it requires about eight tons of average coal to produce it, we arrive at the price of 6s. 3d. per ton—i.e. crediting the coal with half the cost. The capital required to mine eight tons of coal a year in England is difficult to estimate, but it may be taken approximately to be £5, and the capital for plant and machinery to convert it into electricity at £10, making a total of £15. In the case of water-power the average capital cost on the above basis is £40, including water rights (though in exceptionally favoured districts much lower costs are recorded).

From these figures it appears that the average capital required to produce electrical power from coal is less than one half the amount that is required in the case of water-power. The running costs, however, in connection with water-power are much less than those in respect of coal. Another interesting consideration is that the cost of harnessing all the water-power of the world would be about 8000 millions, or equal to the cost of the war to England.

Dowling has estimated the total coal of the world as over seven million million tons, and whether we appraise it at 1s. or more per ton its present and prospective value is prodigious. For instance, at 6s. 3d. per ton it amounts to nearly one hundred times the cost of the war to all the belligerents.

In some foreign countries the capital costs of mining are far below the figures I have taken, and, as coal is transportable over long distances and, generally speaking, electricity is not so at present, therefore it seems probable that capital will in the immediate future flow in increasing quantity to mining operations in foreign countries rather than to the development of the more difficult and costly water-power schemes. When, however, capital becomes more plentiful the lower running costs of water-power will prevail, with the result that it will then be rapidly developed.

As to the possible new sources of power, I have already mentioned molecular energy, but there is another alternative which appears to merit attention.

Bore Hole

In my address to Section G in 1904 I discussed the question of sinking a shaft to a depth of twelve miles, which is about ten times the depth of any shaft in existence. The estimated cost was £5,000,000, and the time required about eighty-five years.

The method of cooling, the air-locks to limit the barometric pressure on the miners, and other precautions were described, and the project appeared feasible. One essential factor has, however, been queried by some persons: Would the rock at the great depth crush in and destroy the shaft? Subsequent to my address, I wrote a letter to *Nature*, suggesting that the question might be tested experimentally. Professor Frank D. Adams, of McGill University, Montreal, acting on the suggestion, has since carried out exhaustive experiments, published in the *Journal of Geology* for February 1912, showing that in limestone a depth of fifteen miles is probably practicable, and that in granite a depth of thirty miles might be reached.

Little is at present known of the earth's interior, except by inference from a study of its surface, upturned strata, shallow shafts, the velocity of transmission of seismic disturbances, its rigidity, and its specific gravity, and it seems reasonable to suggest that some attempt should be made to sink a shaft as deep as may be found practicable and at some locality selected by geologists as the most likely to afford useful information.

When we consider that the estimated cost of sinking a shaft to a depth of twelve miles, at present-day prices, is not much more than the cost of one day of the war to Great Britain alone, the expense seems trivial as compared with the possible knowledge that might be gained by an investigation into this unexplored region of the earth. It might, indeed, prove of inestimable value to Science, and also throw additional light on the internal constitution of the earth in relation to minerals of high specific gravity.

In Italy, at Lardarello, bore holes have been sunk, which discharge large volumes of high-pressure steam, which is being utilised to generate about 10,000 horse-power by turbines. At Solfatara, near Naples, a similar project is on foot to supply power to the great works in the district. It seems, indeed, probable that in volcanic regions a very large amount of power may be, in the future, obtained directly or indirectly by boring into the earth, and that the whole subject merits the most careful consideration.

While on the subject of obtaining power, may I digress for a few moments and describe an interesting phenomenon of a somewhat converse nature—viz. that of intense pressure produced by moderate forces closing up cavities in water.

A Committee was appointed by the Admiralty in 1916 to investigate the cause of the rapid erosion of the propellers of some of the ships doing arduous duties. This was the first time that the problem had been systematically considered. The Committee found that the erosion was due to the intense blows struck upon the blades of the propellers by the nuclei of vacuous cavities closing up against them. Though the pressure bringing the water together was only that of the atmosphere, yet it was proved that at the nucleus 20,000 atmospheres might be produced.

The phenomenon may be described as being analogous to the well-known fact that nearly all the energy of the arm that swings it is concentrated in the tag of a whip. It was shown that when water flowed into a conical tube which had been evacuated a pressure of over 140 tons per square inch was recorded at the apex, which was capable of eroding brass, steel, and in time even the hardest steel. The phenomenon may occur under some conditions in rivers and waterfalls where the velocity exceeds 50 feet per second, and it is probably as great a source of erosion as by the washing down of boulders and pebbles. Then again, when waves beat on a rocky shore, under some conditions, intense hydraulic pressures will occur, quite sufficient of themselves to crush the rock and to open out narrow fissures into caves.

Research

The whole question of the future resources of the Empire is, I venture to think, one which demands the serious attention of all scientists. It should be attacked in a comprehensive manner, and with that insistence which has been so notable in connection with the efforts of British investigators in the past. In such a task, some people might suggest, we need encouragement and assistance from the Government of the country. Surely we have it. As many here know, a great experimental step towards the practical realisation of Solomon's House as prefigured by Francis Bacon in the *New Atlantis* is being made by the Government at the present time. The inception, constitution, and methods of procedure of the Department, which was constituted in 1915, were fully described by Sir Frank Heath in his paper to the Royal Society of Arts last February, and it was there stated by Lord Crewe that, so far as he knew, this was the only country in which a Government Department of Research existed.*

It is obvious that the work of a Department of this kind must be one of gradual development with small beginnings, in order that it may be sound and lasting. The work commenced by assisting a number of researches conducted by scientific and professional societies which were languishing as a result of the war, and grants were also made to the National Physical Laboratory and to the Central School of Pottery at Stoke-on-Trent. The grants for investigation and research for the year 1916–17 totalled £11,055, and for the present year are anticipated to be £93,570. The total income of the National Physical Laboratory in 1913–14 was £43,713, and owing to the great enlargement of the Laboratory the total estimate of the Research Department for this service during the current year is £154,650.

Another important part of the work of the Department has been to foster and to aid financially Associations of the trades for the purpose of

* The Italian Government are now, however, establishing a National Council for Research, and a Bill is before the French Chamber for the establishment of a National Office of Scientific, Industrial, and Agricultural Research and Inventions.

research. Nine of these Associations are already at work; eight more are approved, and will probably be at work within the next two months; and another twelve are in the earlier stage of formation. There are also signs of increased research by individual factories. Whether this is due to the indirect influence of the Research Department or to a change in public opinion and a more general recognition of the importance of scientific industrial research it is difficult to say.

The possibility of the uncontrolled use on the part of a nation of the power which Science has placed within its reach is so great a menace to civilisation* that the ardent wish of all reasonable people is to possess some radical means of prevention through the establishment of some form of wide and powerful control. Has not Science forged the remedy, by making the world a smaller arena for the activities of civilisation, by reducing distance in terms of time? Alliances and unions, which have successfully controlled and stimulated republics of heterogeneous races during the last century, will therefore have become possible on a wider and grander scale, thus uniting all civilised nations in a great League to maintain order, security, and freedom for every individual, and for every State and nation liberty to devote their energies to the controlling of the great forces of Nature for the use and convenience of man, instead of applying them to the killing of each other.

Many of us remember the President's Banner at the Manchester Meeting in 1915, where Science is allegorically represented by a sorrowful figure covering her eyes from the sight of the guns in the foreground. This year Science is represented in her more joyful mien, encouraging the arts and industries. It is to be sincerely hoped that the future will justify our present optimism.

* For instance, it might some day be discovered how to liberate instantaneously the energy in radium, and radium contains 2½ million times the energy of the same weight of T.N.T.

THE RISE OF MOTIVE POWER AND THE WORK OF JOULE

Manchester Literary and Philosophical Society, December 5th, 1922

Discoveries of the fundamental laws of Nature probably result in the greatest intellectual and material benefit to mankind.

Epoch-making discoveries are rare and to the successful pioneers into the realms of Nature the unstinted homage of mankind is appropriately due, and amongst them Joule must undoubtedly rank as one of the greatest.

In this, the second Joule Memorial Lecture, we commemorate his great work and direct our attention to the influence that it has exercised on the progress of the science of motive power.

In the year 1837, when Joule was 19 years of age, the interest in electrical matters, philosophical and practical, was new and keen.

The discovery of Oersted in 1820 that the compass needle is deflected from its usual direction by an electric current parallel to the needle, and the discovery by Sturgeon in 1825 of the soft iron electromagnet, had in 1837 rendered the electric telegraph practical in the hands of Cook and Wheatstone. Ohm had, ten years previously, discovered the relation between the electromotive force of the battery, the resistance of the circuit and the current generated.

Faraday had discovered the current produced in a closed circuit by the motion of a magnet called "magneto-electric induction", and was engaged in his new classical *Experimental Researches in Electricity*, having already shown that the quantity of electricity produced in a battery is proportional to the number of chemical equivalents electrolysed.

In 1836 Sturgeon added a new interest to the subject by his invention of the "Commutator" and his construction of two machines, one the magneto-electric machine, now called the "Dynamo", and the other the electro-magnetic engine, now called the "Motor".

To a young man of Joule's temperament, surroundings and education, the broadening field of matters electrical at this period must have presented extraordinary attractions, and he at once plunged into experiments to try and improve Sturgeon's engine.

There must also, of necessity, have been some additional incentive to Joule to start with the tackling of motive power engines, an incentive which thirteen years before had fired the imagination of Sadi Carnot in France, viz. the great development of the steam engine, taking place in England and then spreading to other countries.

By the year 1774, James Watt had much improved the engine and had reduced its consumption of fuel to one-quarter that of any previous steam engine.

The laws of steam which he had discovered were that the latent heat was nearly constant for any pressure within the range of steam engine practice at that time, and that, consequently, the greater the steam pressure, and the greater the range of expansion, the greater would be the work obtained from a given weight. He had also invented the steam engine indicator and the separate condenser.

James Watt's conclusions appear to have been the result of close and patient reasoning by a mind endowed with extraordinary powers of insight into physical questions, and of drawing sound practical conclusions from numerous experiments devised to throw light on the subject under consideration.

Although Watt himself had a clear perception that higher pressures and greater ratios of expansion would yield a greater proportion of power, nevertheless it seems evident that he never clearly recognised the fundamental principle that heat engines work by taking in a quantity of heat at a high level of temperature, and rejecting it (in lesser quantity) at a lower level, the difference between the two quantities representing the external work done.

In fact, in spite of the stir caused by Count Rumford's famous experiments on frictional heat (about 1798) the full significance of that drop in heat in the engine was not appreciated until many years later.

In 1824, Sadi Carnot published a little book entitled *Réflexions sur la Puissance Motrice du Feu*, which at that time had only a limited circulation and passed almost without notice. Carnot died in 1832, at the early age of 36, but in this little book he wrote:

The discovery of the steam engine owed its birth, like most human inventions, to rude attempts which have been attributed to different persons while the real author is not clearly known. If the honour of a discovery belongs to the nation in which it has acquired its growth and all its developments, this honour cannot here be refused to England. Savery, Newcomen, Smeaton, the famous Watt, Woolf, Trevithick, and some other English Engineers, were the veritable creators of the steam engine. It has acquired at their hands its successive degrees of improvement. Finally, it is natural that an invention should have its birth, and especially its development, and be perfected in that place where its want is most strongly felt.

Notwithstanding the work of all kinds done by steam engines, notwithstanding the satisfactory condition to which they have been brought to-day, their theory is very little understood, and the attempts to improve them are still directed almost by chance. The question has often been raised whether the motive power of heat is unbounded, whether the possible improvements in steam engines have an assignable limit.

Carnot seems to have been the first to realise the idea of the pressure-temperature cycle of the working substance in a heat engine, whose function it is to produce motive power by taking in heat at a high level of temperature and rejecting it at a lower level of temperature, for he says "wherever there exists a difference of temperature motive power can be produced".

Carnot, however, failed to recognise that heat disappeared in the process in an amount equivalent to the external work done. This disappearance of heat, it should be remembered, was small in steam engines constructed up to Carnot's time, viz. 1824—not exceeding 5 per cent. of the total as compared with 20 to 30 per cent. in the best modern engines—and even if such loss had been noticed it might easily have been attributed to errors in observation. Notwithstanding this fact, Carnot's ideal diagrams with isothermal and adiabatic lines represented very closely the action of the steam in engines of that period, and the ideal indicator card of a steam engine with a two- or three-fold ratio of expansion in the cylinder.

Carnot's work led him to the same conclusion as Watt had reached by somewhat different reasoning, that the greater the fall of temperature between the boiler and the condenser, the greater is the resulting economy of the engine. But Carnot had gone further and generalised the conclusion that the ideal efficiency of a reversible engine is independent of the nature of the working substance, and depends only upon the temperature of the latter when it is absorbing heat and when it is rejecting it.

It was not until 1849, or twenty-five years after Carnot's treatise, that the work of Joule definitely established the indestructibility of energy and the mechanical equivalent of heat, also that heat is a form of energy—a mode of motion in matter. Joule went on to suggest the existence of the absolute zero of temperature. He argued thus: since the pressure of a gas in a closed vessel is proportional to its temperature, and since this has been proved by experiment to be proportional to the internal energy, or heat, in the gas, it follows as a simple deduction that the absolute zero of temperature corresponds to the zero of pressure as indicated on the gas-thermometer, or 493° F. below the freezing point of water.

Sir William Thomson in the next year deduced the absolute zero of temperature from Carnot's cycle; both Joule and Thomson reached the same numerical value of the absolute zero by these different methods. The correction of the Carnot cycle and diagrams followed. Subsequently these questions received mathematical treatment at the hands of Clausius, Sir William Thomson, Rankine and others, and the introduction of the idea of "entropy" enabled the heat-engine cycle of operations to be presented in such a way that the disappearance of a certain quantity of heat became an obvious necessity.

Thus, to summarise, the discovery of the laws of thermodynamics resulted from the work of many persons. The work of James Watt forcibly directed attention to the subject. In Carnot, the initial stimulus was awakened by watching steam engines in France, the development of which he attributes to English engineers, and especially to James Watt. Then Joule's work followed and established another great law, which, together with Carnot's, supplied the ideas and the data which the leaders of Science, having reconciled and placed in logical order, gave to the world as tools of great power, suitable for general use by other scientists, physicists and engineers.

Joule at Manchester was much closer geographically to the development of the steam engine than Carnot in Paris, but though a brilliant experimenter in the laboratory, and, as Osborne Reynolds tells us, with all the skill of an engineer and mechanician, he chose, as was perhaps natural, under the circumstances, the more unknown and to him attractive field of endeavouring to improve the means of obtaining power from the electromagnetic engine worked by consumption of zinc in a battery, and although its results were disappointing it proved an education in itself and led him on to his great life's work.

Let me revert for a few moments to Joule's early work, on the electro-magnetic engine, which he commenced when 19 years of age and continued for three years.

He found to his disappointment that the speed of his engine was limited and realised the cause to be self-induction. He had discovered the fundamental law of attraction to be proportional to the square of the ampere turns; but not yet the law of the conservation of energy, and he was dazzled by the possibility of practical results. He saw no objection to perpetual motion for he had then no idea as to the relation between mechanical work and its equivalent in heat. He does not give up the research but goes more minutely into it. He measures the energy of the electric current passing to the motor, the electrical and mechanical losses in the motor and even the heat evolved in the iron core of the armature, by running it in the electric field without its windings. Thus from the balance sheet which he constructs he arrives at the figure of 838 foot pounds as the equivalent of one thermal unit; a wonderfully close approximation to the true figure of 772* having regard to the complexity and difficulties of this method of attacking the problem. In fact his arrangements and precautions to secure accuracy were worthy of the most highly skilled modern pioneer in electrical apparatus and machinery. It will be interesting to quote a few lines from the lecture delivered by Joule in 1865 at Greenock in commemoration of the birth of James Watt and about fifteen years after the general acceptance of the results of Joule's great work in establishing the mechanical equivalent of heat and the laws of thermo-dynamics:

* Now generally accepted as 778 on the British system.

We have met on the anniversary of the birth of the greatest mechanician that this or any other country has ever produced—of one who not only commenced and carried on a great work, but who bequeathed it to posterity in a perfect state. The officers of this Society have thought it fitting that, on the occasion thus presented to us, we should reflect on some of those facts of Physical Science which are intimately connected with the discoveries of Watt, and which give a rational explanation of the operation of that wonderful machine which will for all time be associated with his name.

Joule then proceeds to recite the growth of the knowledge of heat, as follows:

The Ancients recognised heat as the source of life—and the main instrument by which Nature is carried on. This conception was improved by Bacon, Newton and other illustrious men who adorned the seventeenth century. Heat was considered by them as a motion among the particles of matter. "Heat", says Locke, "is a very brisk agitation of the insensible parts of the object, which produces in us that sensation from which we denominate the object hot; so what in our sensation is heat, in the object is nothing but motion." During the same epoch Newton, in his *Principia* laid down the axioms and laws of motion, and announced the principle of the conservation of energy. So that the solid foundation of the Science of heat was laid two centuries ago.

Now let us consider what is the influence that the knowledge of the laws of thermodynamics has exercised on the progress of Science and the Arts. The influence extends over many fields, including Thermo-Chemistry, Metallurgy, Electricity, Biology, Astronomy, Motive Power and many others, but this evening we must restrict our attention to Motive Power.

It is somewhat strange that the earliest recorded attempt to obtain motive power from heat should have been by means of a steam wheel or Reaction Turbine, and not, as one might have expected, a piston engine. In Hero's turbine (about 130 B.C.) the steam was introduced at the axis, and issuing from nozzles at the extremity of hollow arms, drove the wheel by reaction.

James Watt had under consideration the probable performance of the Hero type of engine in 1784*, for in a letter to his partner, Matthew Boulton, he discusses at some length the prospects of success of Baron von Kempelen's engine in Hungary, working on the Hero principle, as a possible rival to the piston engine. In this letter he states approximately the correct velocity of issue of steam from a non-divergent nozzle, but he estimates the force of the reaction of the jet incorrectly, assigning to it only one half the true amount, and his letter to Boulton ends as follows: "In short, without God makes it possible for things to move 1000 feet per second, it cannot do much harm".

The next heat engine mentioned in history is also a steam turbine—

* *Some Unpublished Letters of James Watt.* Published by the Institution of Mechanical Engineers. 8vo., London, 1915.

Branca's steam wheel in the year A.D. 1629. The action is on the impulse principle, the steam nozzle being fixed and playing upon buckets.

Both the Hero and the Branca types were undoubtedly capable of practical development for commercial uses, but the state of the mechanical Arts was not then sufficiently advanced to make possible the construction of such apparatus in an efficient form or to apply the power effectively.

Not until a century after Branca, in about the year 1705, does the first reciprocating piston engine make its appearance in association with the names of the Marquis of Worcester, Papin, Savery, Newcomen, Smeaton, and, half a century later, James Watt. Further, not until the year 1837, at a time when the Watt engine had attained extensive use, do we find attention again directed to the steam wheel, when several were constructed on Hero's principle by Dr Wilson of Greenock, and by Avery of Syracuse for driving circular saws and cotton gins.

The tubular arms of these wheels were 2 ft. 6 in. long and of oval or stream-line section. The nozzles were non-divergent and the peripheral speed 880 feet per second. The steam consumption is not recorded, but we may calculate that it was somewhat inferior to that of an ordinary piston engine.

I now come to a period when the great laws of thermodynamics were recognised by physicists and engineers. To test the possibilities of the Hero turbine I constructed one about thirty years ago, with divergent nozzles. When tested with steam at 100 pounds pressure, and with the jets revolving at a peripheral speed of 800 feet per second in a chamber and with a good vacuum, it was found to give about the same efficiency as a good single cylinder reciprocating engine. This type, however, though of historical interest, does not lend itself to higher refinements as readily as Branca's turbine which had lain dormant until 1883 when Dr De Laval of Stockholm commenced to develop it, and adopting the divergent nozzle in 1887 realised economies equal to that of the ordinary piston engine. The diverging nozzle was, in fact, suggested by Pilbrow and others half a century before, but De Laval actually proved its practical advantage and utilised it in commercial engines. The theory of the divergent nozzle was first treated mathematically, I believe, by the late John Perry some years afterwards. At the same period De Laval also introduced double helical gearing, which had previously been used only in clocks and small apparatus, and proved that it was a practical proposition for transmitting considerable powers whilst reducing the then phenomenal speeds of rotation of his turbine down to suitable speeds for ordinary commercial uses. This form of gearing has in recent years proved to be a very important factor in turbine development, both at sea and on land.

Recurring to the De Laval turbine, the diverging nozzle would seem at first sight to be an ideal instrument for converting the expansive energy of steam into kinetic energy of linear motion with a high degree of efficiency, but this is not so. Certain inherent losses exist, reaching nearly 20 per cent. in the nozzle when dealing with high ratios of expansion; and although the subject has received much attention at the hands of experimenters and scientists, this loss has not been adequately explained either by the assumption of supercooling or by the dissipation of energy by sound waves. Incidentally it will be remembered that loss of energy by sound was shown by Joule to exist in his celebrated experiments on the heat equivalent of mechanical work in gases when he caused them to flow through a tap from a vessel at high pressure into another at lower pressure.

If there were no losses in the steam jet, the linear velocity attained by dry-saturated steam initially at 100 pounds per square inch, flowing into a vessel at an absolute pressure of 1 pound per square inch, would be 3800 feet per second, and the surface velocity of the bucket-wheel for maximum efficiency should then be about 1700 feet per second. In practice, however, the highest linear velocity of bucket attainable is, because of high stresses and windage losses, only about 1200 feet per second, so that the proper velocity ratio for high efficiency cannot be realised, and consequently there is the loss by shock on the blades to be added to the loss in the jet. For these reasons, turbines which deal with the steam in a single pressure-drop or expansion are much less efficient than those in which the steam is caused to flow through a large number of turbines in series gradually increasing in size, with a small pressure-drop at each turbine. All large turbines are now built on this principle, which is called "Pressure Compounding", a principle which was first proposed by Wilson and others about 1846, although a turbine does not appear to have been constructed on this principle until nearly forty years later, when in 1884 I commenced to work at this interesting subject.

Attention had been almost solely concentrated on the development of the piston engine up to this period with the result that, assisted by the knowledge of the laws of thermodynamics, great progress had been made by the raising of the pressure and the provision of greater ratios of expansion; this was achieved by the compounding, tripling and quadrupling of cylinders of increasing capacity, in series on the steam, and in some cases by the use of superheat. It had at length been found that the maximum efficiency was reached with about 16-fold expansion by volume in the engine. Beyond this there appeared to be no practical gain because of the disproportionate increase in frictional losses, cost of construction and maintenance. In other words, by 1880, the reciprocating engine had nearly reached its maximum economy.

In commencing to work on the steam turbine in 1884, it appeared that as the laws for the flow of steam through orifices, under small differences of head, were known to correspond closely with those for the flow of water, and that the efficiency of water turbines was known to be from 70 to 80 per cent., the safe course was to adopt the water turbine as the basis of design for the steam turbine. In other words, it seemed reasonable to suppose that, if the total drop of pressure in a steam turbine were to be divided up into a large number of small stages, and an elemental turbine like a water turbine placed at each stage, then each individual turbine of the series ought to give an efficiency similar to that of the water turbine, and that for the whole aggregate of turbines a high efficiency would result; further, that the speed of revolution for maximum efficiency could be easily reached; the first turbine was therefore pressure compounded.

It soon became evident that the mechanical difficulties which had been partially solved in the case of small turbines could be more completely solved in larger sizes; the clearances and consequent leakages of steam could be reduced in greater proportion to the size and output of the engine. The curvature of the blades and their heights could be more accurately adjusted so that each turbine of the series would be more correctly proportioned to cope with the density and drop in pressure, corresponding with its position in the series, and that, in short, a large turbine could be constructed to expand the steam continuously from boiler pressure right down to the lowest vapour pressure attainable in the condenser, with an overall efficiency from 70 to 80 per cent. of a perfect engine working under the same steam conditions. In other words, it seemed probable that higher efficiencies would be realised by a pressure compounded turbine than by any form of piston engine.

This has ultimately been found to be the case in large compound turbines. Thus, in a certain modern 10,000 kilowatt turbo-alternator supplied with steam at 250 pounds per square inch, superheated to 700° F. and exhausting into a condenser vacuum of 29 inches (or an absolute pressure of $\frac{1}{2}$ pound per square inch), the steam is expanded to one five-hundredth of its original pressure, and under these conditions one quarter of the heat energy contained in the steam is converted into electrical energy. Further, after allowing for the losses in the boiler plant and for the power used for operating the auxiliaries, about 20 per cent. of the heat in the coal is converted into electrical energy.

It seems, moreover, that in still larger turbine plants, with a steam pressure of 500 pounds per square inch, superheated to 700° F., and with reheating to 700° F., also with feed-water heating in stages by means of steam withdrawn from the turbine, and preheating of the air to the boilers by heaters taking heat from the flue gases after the economisers, it is probable that a thermal efficiency of over 32 per cent., and an overall thermal efficiency from coal to electrical energy, of over 27 per cent., may be reached.

The turbine takes in heat chiefly at the temperature of the boiler and also in much lesser quantity at the temperature of the superheater, whilst it rejects heat at the temperature of the vapour in the condenser. Thus, in very large turbines, the temperature of intake of heat may average 450° F., and the temperature of rejection 70° F.

Within the last half century another method of obtaining motive power from heat has come largely into use. The history of the internal combustion engine commences with Huygens' proposed gunpowder engine in 1680, followed by the proposal to use oil or gas, with compression before ignition, about the year 1838. Not, however, until 1860, when the fundamental bearing of the laws of thermodynamics and the advantage of compression before ignition were recognised, did the great inventors Bennett, Otto, Clerk, Diesel and many others achieve developments which have led to the attainment of the very high efficiency of the present-day internal combustion engine.

In all such modern engines the principle is, broadly speaking, the same, in that the working fluid is first raised to a high temperature by adiabatic compression; heat is then added by combustion of a portion of the working fluid, the products are then cooled by adiabatic expansion, and finally heat is rejected at the temperature of the exhaust gases. Thus we see that the cycle approximates roughly to the Carnot cycle and that the theoretical efficiency depends mainly upon the ratio of compression. Mechanical difficulties and the increased cost and weight of the engine in proportion to the power produced are the factors which limit the degree of compression that is commercially attainable.

Nevertheless, when making a comparison between the heat in the fuel and the work done on the brake, very high thermal efficiencies have been recorded during special tests of large gas and Diesel engines, for example:

Large suction gas engines   24·2 per cent. brake thermal efficiency.
Large gas and Diesel engines 32·5   ,,   ,,

To make these figures comparable with those for the steam turbine which I have given in terms of electrical output, they must be multiplied by the efficiency of the dynamo, taken at 94 per cent., and debited with the cost of lubricating oil for the cylinders, which I have taken as 10 per cent. of the cost of the fuel-oil. We then have:

Large gas and Diesel engines   27·5 per cent.
Large anthracite suction gas engines 20·5   ,,
Very large steam turbines   27   ,,
Large steam turbines   21   ,,

It is beyond the scope of this address to enter upon the relative costs of these two principal means of deriving motive power from fuel, but I may add that for large powers the steam turbine at present holds the field almost exclusively on land and at sea, while the internal combustion engine has superseded steam to a large extent for moderate and small powers on land and is to some extent a similar rival at sea.

The internal combustion turbine has received in recent years the attention of many inventors and experimentalists, and some have argued that, following the analogy of the steam turbine having largely superseded the piston steam engine, the gas turbine ought therefore to supersede the gas engine. The analogy, however, is at fault, for the steam turbine and steam engine work between limits of temperature within which metals will stand without melting or serious oxidation, whereas the gas turbine and gas engine have to deal with a working fluid at temperatures far above the melting point of the metals of which they are constructed.

Further, in the steam turbine and the steam engine the metal is nearly at the same temperature as the working fluid, whereas in the gas engine and gas turbine the metal must be kept at a much lower temperature than the working fluid, otherwise it would melt.

All gas turbines, of whatever type, involve the passage through the jets and blades of highly heated gases at great velocities, and if it is attempted to reduce such high temperatures by dilution with gases of lower temperature, or by the injection of water or steam, the thermal efficiency of the engine is much diminished.

Recently it has been claimed by Holzwarth that the blades of his gas turbines, which are made of mild steel, have successfully withstood the high temperatures of the products issuing from the jets of his explosion chambers. He also claims to have attained a thermal efficiency not much inferior to that of the piston gas engine, but there seems to be some doubt on these points.

Contrasting internal combustion turbines generally, which involve hot gases moving at high velocities, with the internal combustion piston engine with the hot gases contained in bulk in the cylinder, and, except for a little turbulence in the charge, quiescent and not allowed to flow at velocity through the exhaust valve till they have been cooled by adiabatic expansion, and realising how much more favourably the metal of the cylinder and piston is situated than are the blades in a turbine, there seems to be little question that as regards durability and efficiency the piston is a preferable means of expanding the hot gases and abstracting their energy.

In the time at our disposal it has not been possible to do more than draw attention to some of the more interesting periods and events in the development of motive power. We have noted that in the days of Watt the consumption of fuel in proportion to power was reduced fourfold, and in the days of Joule, and after, it has again been reduced by another fourfold, making 16-fold since the days of the Newcomen engine.

Another fourfold reduction would bring us to an engine of 100 per cent. overall efficiency, an ideal which is probably beyond our fondest dreams of attainment, unless indeed our minds may be dazzled by the vision of working in fields where the converse of Joule's mechanical equivalent of heat operates and the harnessing of the energy in matter is possible.

In conclusion, I wish to thank your Society for the great honour it has done me in allowing me to address you, however imperfectly, on a subject so wide, important and interesting.

PRESIDENTIAL ADDRESS

The Institute of Physics, London, May 26th, 1924

Two years have passed since the Inaugural Meeting of this Institute, on which occasion speeches were delivered explaining the objects which the promoters had in view and explaining the work of the physicist in pure science, scientific research and the part played by science in the arts and manufactures.

Perhaps it may be interesting to recall a few instances from past history bearing upon the association of physics with the advancements in the arts and sciences.

James Watt, whose great work was in engineering, was by it led to discover more about the properties of steam, which were imperfectly known at the time; he had many friends among physicists and chemists, but their knowledge of the subject in which he was so deeply interested did not satisfy him. He, therefore, experimented himself and drew his own conclusions which, though approximate, were yet sufficiently correct to guide him to his epoch-making achievements. He would now probably be described as an engineering physicist. Further, as regards the business side of James Watt, it appears doubtful whether his work would have resulted in great epoch-making results, had he not early on entered into partnership with a sympathetic business capitalist—Matthew Boulton. Then again, though less apparent, there can be no doubt that the work of Watt influenced physicists to direct their attention to the accurate determination of the properties and laws of steam—and not long afterwards to the founding of the great laws of thermodynamics; for Joule's great work was in physics, but we find him first trying to improve an electromagnetic engine worked by the electrolytic consumption of zinc, which he thought would supersede the steam engine of Watt as a prime mover; in this work he showed the skill of the best electrical and mechanical engineer. In this attempt he, however, failed, but his experiments had the result of leading him to the great discovery of the mechanical equivalent of heat, and the absolute zero of temperature, and it would seem probable that, had he not at first attempted an impossible physical and electrical engineering problem, he might never have determined the mechanical equivalent of heat. Again, Sadi Carnot was drawn towards his great work on thermodynamics by watching the Watt steam engines, some of which were then working in France, and by pondering on the function of the steam as the working fluid, how it acted within the cylinder, what was the heat cycle, and what was the relation between the heat of the fuel burnt under the boiler and the work done. He pondered over these things and eventually evolved the solution: the "Carnot Cycle."

He was not, however, fully aware of the mechanical equivalent of heat or of the absolute zero of temperature; his solution was, in consequence, only approximate, but the accurate solution followed later and the reconciliation of the discoveries of Joule and Carnot was subsequently developed by Clausius, Kelvin, Rankine, and others.

Thus were the great and fundamental laws of thermodynamics discovered.

Reviewing more recent times—the late Lord Armstrong,* known chiefly as the introducer and manufacturer of hydraulic machinery and guns, was a physicist in methods of thought and methods of investigation of engineering problems and principles. He was not a mathematician, but knew well how to direct the mathematical and experimental treatment of a mechanical problem, and of the thermodynamic problems of guns. He was more successful than Whitworth, chiefly because he was a better physicist and had a better faculty for dealing with and applying the fundamental laws of physics. He was an admirable organiser of team work, and an astute man of business with quick observation of the physical conditions of things. On one occasion, by chance, he observed that a tie bar in the roof of the boiler house emitted electric sparks; he found that a leak of steam from an adjacent pipe was playing upon the rod, and that the roof being of wood, the tie bar became electrically charged. He then carried out experiments and found that jets made of ivory of such a shape as to cause a maximum amount of friction to the issuing steam gave the best results. He thus invented his hydro-electric machine for generating high tension electricity from saturated steam.

When Sir Isaac Newton was asked how he made his discoveries, he replied " by always thinking unto them, I keep the subject constantly before me, and wait till the first dawnings open slowly by little and little, into a full and clear light." Sir Oliver Lodge here remarks: "That is the way—quiet, steady, continuous thinking, uninterrupted and unharassed brooding. Much may be done under those conditions—all the best thinking work of the world has been thus done," and Newton adds: "If I have done the public any service this way, it is due to nothing but industry and patient thought."

Faraday in his lecture on the Forms of Matter to the City Philosophical Society in 1819 when he was twenty-seven years of age said:

The disagreeable and uneasy sensation produced by incertitude will always induce a man to sacrifice a slight degree of probability to the pleasure and ease

---

* [Lord Armstrong was a young solicitor when he invented his hydraulic engine as the result of an observation on an inefficient water-mill made during a fishing expedition in Yorkshire. Ed.]

of resting on a decided opinion, and where the evidence of a thing is not quite perfect, the deficiency will be easily supplied by desire and imagination. The efforts a man makes to obtain a knowledge of nature's secrets merit, he thinks, their object for their reward; and though he may, and in many cases must, fail of obtaining his desires, he seldom thinks himself unsuccessful, but substitutes the whisperings of his own fancy for the revelations of the goddess.

When Newton, at the age of twenty-three, discovered that the motions of the solar system were due to the action of a central force, varying inversely with the square of the distance, his theory was based on Kepler's Laws and was clear and certain, and he might have rushed into premature publication after our nineteenth-century fashion, but that was not his method.

He tried whether the force of gravity would serve and he found that it was 15 per cent. too strong, owing to the incorrect numerical data as known at that time, and he thereupon laid aside all thought of the matter for sixteen years, when new and more accurate determinations of gravity were available which enabled theory to be reconciled with observation.

The methods followed by Newton were daring but restrained by caution; as indeed were those of Faraday, but Faraday's warning seems to indicate that other methods of thought and procedure were prevalent in his day, which indeed seem to have become still more common in our time, in science—physics and chemistry—but more especially so among pioneers in engineering and in the arts generally. So much is this the case that the name of inventor has become distasteful to those who have done really useful work, the best work in engineering and in the arts, or to those who as Newton has said "have done the public any service this way."

There is undoubtedly a strong tendency induced by constantly thinking about one subject—to view it with distorted perspective, and incorrect magnification, thus leading to wrong and unjustifiable conclusions, so well described by Faraday.

In this connection it is interesting to consider the mental attitude and method of the skilled and really useful inventor. The inventor must have a wide and intimate acquaintance with the properties of all the materials with which he has to work, and by long and patient thought must endeavour to raise in his mind a picture of the ideal combination he is seeking to form out of the materials at his disposal. In the search he will picture in his mind edifice after edifice, combination after combination, machine after machine, and examine how far they conform to the known laws of architecture, physics, chemistry and engineering; some may be seen at once to contravene those laws and are discarded, while others more plausible will go to the touchstone of mathematical calculation and tentative experiment; and, should this reveal no imperfection or palpable defect, then to more rigorous or costly research and experiment. In short the skilled inventor marshals his ideas and following the procedure of Nature puts them through a process of "Natural Selection" with the object of obtaining the fittest solution of his problem. The testing out of any discovery or invention is probably, in general, the most difficult and the most important part of the work, a work in which a knowledge of physics is of inestimable value: in most cases it is a matter of hard work and patience.

That some scientists but many inventors are prone to jump to conclusions too readily, seems to be principally due to the lack of scientific knowledge and methods of thought among inventors. Invention is, in almost all cases, the result of the work of many persons and very rarely that of one man. To quote from Sir Dugald Clerk's Truman Wood Lecture, at the Royal Society of Arts in 1917: "Sir Berkeley Moynihan* when reviewing the great advances that had been made in surgery in recent times summed up as follows: A discovery is rarely the work of one mind. It is one observation added to another that makes the supersaturated solution from which the crystal of truth at last precipitates."

Within the last three-quarters of a century the fields of science and the arts have so much developed in extent, complexity and refinement that it has become impossible for a single brain to grasp effectively more than a small proportion of those wide fields with any prospect of making an advance, and it has become necessary to specialise, to collaborate and to delegate work, in order to realise success.

The increased complexity and refinement in all branches of research, in applied mathematics, physics, astronomy, engineering, chemistry, textiles, ceramics, glass making, involve more capital in the shape of trained brains and apparatus of all kinds. The physicist wants more complex apparatus, the astronomer larger and more highly developed telescopes, clocks, spectroscopes, cameras, interferometers, etc. Engineering requires an increasing quantity of apparatus of all kinds and occasionally experiments have to be made on a large scale and at great cost and, in other words, more money is required. As Sir Arthur Keith has well said: "Capital is as old as Nature, for consider the hen's egg—the yolk is the capital and without it the little entities in the egg could not work to build up the chicken." As research advances in complexity more capital is required to carry on the advance.

The hopelessness of making progress in the absence of data and an intimate knowledge of a subject was deeply impressed on the minds of those who served on the Boards of Invention and Research during the war, and who received at one period more than 2000 inventions and suggestions per week, of which only a minute fraction of 1 per cent. were in any way helpful, while scarcely any were of practical value. The immense wastage of mental effort appeared to them deplorable.

* Now Baron Moynihan of Leeds.

In manufacturing works many years ago, the chief men who owned or controlled them usually conducted such experimental work as was considered necessary, and among them were to be found many good physicists; to mention two instances only, Lord Armstrong in engineering to whom I have already alluded, and Sir Lowthian Bell in the metallurgy of iron, though many others might be cited in these and other industries. These men found time for research by delegating routine work to their colleagues or subordinates, and they were thus enabled to give a portion of their time, and undisturbed attention, to research on the problems of manufacture and improvements of processes in which they were interested. Though this procedure is still common, there has been a slow but universal tendency in large and even in works of moderate size to establish research departments under and in close collaboration with the management. There may be one or more such departments in touch with each other, in which case they collaborate and afford each other mutual assistance and advice. Equipped with the best instruments and with a highly trained staff they form a component part in the organisation of the works, and are of primary importance in many industries. Care must, however, be taken that there is no loss of collaboration between the management and the research workers, and to this end it is desirable that the heads themselves should have some training in physics, or evince a due appreciation of the advantages to be derived from research, and also have some capacity for directing, and for sympathetic collaboration with, the research workers.

The development of the steam turbine, to which I have directed much attention, may perhaps be taken as an illustration of a research in one class of engineering, and it may probably also be taken as representative of researches in many other lines of manufacture. At the outset the collection of new data was obviously required before the general line of advance could be determined. Some preliminary experiments were made with high speed shafts and bearings, but in order to complete the data a small turbine coupled to a high speed dynamo of primitive design was made; the calculated stresses due to centrifugal force, the laws governing the flow of steam and data from dynamos as approximately known at that time were taken into account. This machine was tested on very similar lines to those followed by Joule when trying to improve the electromagnetic engine of Sturgeon. The constants for the flow of steam, the loss by friction in bearings at high surface speeds, the hysteresis and eddy current losses in armature core, conductors, and binding wire at abnormally high speeds were approximately investigated. Higher mathematics were not employed in this work, but were used much later to coordinate the accumulated data and forecast the effect of small improvements and refinements which have, in recent years, considerably increased the thermal efficiency of the turbine; as a matter of fact, it does not now appear that the use of higher mathematics in the earlier stages of development would have been helpful; the accumulation of sufficiently accurate data to have enabled them to have been practically and usefully applied would have been at that time an additional burthen and hindrance to progress. This, however, does not imply that a mathematical and physical training is not of very great value, for the two men who directed the work had passed mathematical and physical courses at universities, as well as being trained engineers. All that it is intended to emphasise is that higher mathematics should be used in its proper place, and in engineering and pioneer work chiefly to consolidate the rear, assist the communications, but seldom to lead in the advance.

On the other hand, the coming of the steam turbine has had the effect of stimulating research in certain directions, notably in the dynamics of rotating shafts, the law of flow of saturated and superheated steam through jets, and the frictional resistances to flow through passages and over surfaces; also to super-cooling and other phenomena.

In conclusion it may be desirable to consider briefly the position of the manufacturers of scientific instruments and materials in this country, an industry which has never been of a very lucrative character. A century or more ago it was a very small industry assisted to some extent by the patronage of the monied classes; an industry in which the owners of the workshops and the artisans in many instances worked more because of their love of science and scientific achievement than as a means of obtaining a living.

The industry grew considerably during the century before the war and during the war was rightly placed on the list of Key Industries. During the war the output expanded many fold in volume; in some cases assisted by government subsidy and guaranteed orders at remunerative prices, but more generally by means of capital expenditure, prompted by patriotism, of the industry itself. After the war all government subsidies ceased, and the industry now finds itself with works and equipment far too large for normal post-war output, and consequently burthened with abnormally heavy standing charges.

The slump in trade generally, the diminished purchasing power of the public and more especially of the monied classes, also the disposal of surplus war material have affected this industry more than perhaps any other, with the possible exception of the shipbuilding industry.

There is a maxim in commercial business that export of a manufactured article cannot exist unless there is a flourishing home industry in that article. In considering this subject it should be remembered that industries in England depend largely upon exports, and that England has, for a long period, been called the Workshop of the World; but that this unique position has now been modified and powerful rivals have sprung up. This change seems to be most noticeable in some of the so-called Key Industries and, perhaps, especially so in the scientific instrument industry. It is also to be traced in the production of a number of materials of importance to the scientific professions and used in laboratories and scientific workshops. Many instruments which are scientific in their origin are of interest to the general public. The scientific instrument industry is relatively a small one, the nominal capital involved being perhaps in the order of £2,000,000; at present market prices considerably less. Though it is scheduled for protection in the Key Industries Act yet, beyond the placing of an import tariff of $33\frac{1}{3}$ per cent. in respect of such goods, the industry has, so far as I am aware, received no financial help from the State except the Government's contribution to research. (Moreover, the Dominions, in some instances, have of late been purchasing directly from foreign countries. In the case of Canada a large number of optical instruments used there have come from the United States of America.)

It should be remembered that small industries and small works are at a disadvantage, in many respects, in comparison with large industries and large works, because of the small funds at their disposal and their consequent inability to afford an adequate business organisation or special experts for safeguarding their industrial interests. Further, I have been led to think from my short experience in the instrument manufacturing business that perhaps a too-exclusive individualism, involving unnecessary overlapping and lack of interchange of knowledge among the various companies, may have been more pronounced than in some other lines of manufacture.

I turn now to another aspect of the position which seems to me to have considerably retarded the manufacturing development and upon which the experiences of the war have cast light.

For a long time before the war there had arisen a practice among physicists and others equipping their laboratories with instruments, materials and chemicals chiefly imported from abroad, with the result that not only was the home industry impoverished because of lack of orders, but the industry was, to some extent, put out of touch with the requirements of the home market. It thus became less capable of keeping abreast of the most recent advances and of taking a lead by anticipating the wishes and aspirations of the physicists and other scientific workers. As a result the home industry grew less able to compete with the flourishing industrial concerns in foreign countries not so handicapped. On the other hand, it has been alleged by the physicists on their side that because of the limited funds at their disposal they were obliged to buy in the cheapest market provided the article was equally good.

This, to some extent, is true, but at the same time the situation has been largely brought about by the practice I have mentioned; for in spite of the higher cost of British labour it is certain, if the home manufacturers had received more encouragement and more sympathetic help from the buyer, that under these more favourable conditions they would have been able to have provided themselves with better manufacturing equipment, and a more skilled staff, and to have manufactured as cheaply and as well as the foreigner, under a level exchange.

The great advantage to the buyer of having a well-equipped home manufacturer with whom he can consult and collaborate, and who could carry out on the spot his ideas, has not, I think, hitherto been sufficiently recognised in this country.

Now what is the remedy? I think you will agree with me when I say that it is more sympathy between the buyer and the manufacturer and more combination and a less exclusive and self-regarding individualism among the manufacturers; that each firm should concentrate on the few articles that are best suited to its equipment and the abilities of its staff, and develop those with vigour and, in the case of some articles in large demand, adopt mass production.

On behalf of users of scientific instruments I would venture on a word of caution to manufacturers who are looking for cordial support of their industry by their own public. It is that they should bear in mind the need of constant investigation and experiment to bring about improvements in accuracy and adaptability of their instruments, and that they should be ready to show an interest in the demands of the public and do their utmost to meet those demands, even if it means the re-casting of their stereotyped methods. At the present time when British manufacturers of types of instruments which have been previously bought abroad are earnestly endeavouring to ascertain and to meet every wish of the scientific worker, it seems rather an unnecessary caution; but those of us who can remember the time when there was a monopoly in the production of certain instruments in this country recall the "take it or leave it" attitude with which their particular requests were met and realise how well many of our own manufacturers had themselves paved the way for foreign competition.

To the scientific users I would also make an appeal. It is not so widely known as it should be, even among British scientific workers, that if, in some classes, British scientific instruments may fall somewhat below the standard of foreign instruments, in others they are unquestionably superior.

A statement—originally, it may be, true in substance—is made that a certain type of foreign instrument is superior to any other; the statement grows to a legend and lives long after changes and developments have rendered it false or, at least, misleading; and the British instrument has to overcome much inertia of prejudice and fashion before it can secure the recognition which its merits deserve. It would be well if the leading scientific users of instruments would review, from time to time, their judgments of the quality and performance of instruments, so that improvements may receive early recognition and the British manufacturer not be prejudiced by a belated preference for foreign instruments.

STEAM TURBINES

The First World Power Conference, Section D, July 4th, 1924

I. THE STEAM TURBINE ON LAND AND AT SEA

The modern compound steam turbine commenced its development for land purposes in the year 1884; it gradually and steadily grew in size, and though at first it earned the name of "steam eater", it soon improved and in a few years (by 1892) was able to rival the best reciprocating engines in economy of steam and coal for the generation of electricity.

At this period (1894 to 1897) it also entered into the field of marine propulsion.

On land, turbo-electric generating units have been developed to very large sizes, ranging up to 60,000 kilowatts, and in all large electric power stations using fuel, steam turbines now provide the motive power. Some of these stations reach a capacity of 500,000 kilowatts, feeding into widespread distribution systems and covering large areas.

Land turbines, having developed under less restrained conditions than those ruling on board ship, have led the way in economy of steam, and every means has been explored to render large power plants for the generation of electricity as thermally efficient as possible.

Considerable advance in thermal efficiency is being made at the present time. The structural design and the blading of the turbine having been developed to such an extent as to give nearly the maximum efficiency of conversion of the available energy of the admitted steam, attention has been turned to increasing this available energy by extension of the thermodynamic cycle. This involves progress in the performance of boilers and superheaters. By the use of higher steam pressures and temperatures together with the lowest possible exhaust pressure, and by other devices to be described later, it is expected in the future to reach an overall thermal efficiency (from fuel to electricity) of 30 per cent., which is not inferior to the best results attained by oil engines, even neglecting the cost of oil for cylinder lubrication.

On the electrical side improvements have been made in the design of large high-speed alternators. The high-speed alternator of large power, although having no great advantage as regards efficiency over the low-speed type, is lighter for a given output, and occupies less engine room space; both important advantages. Moreover, it can be directly coupled to its steam turbine running at high revolutions, which is conducive to the highest overall efficiency.

High-speed alternators have a further advantage in that they can as a rule, even in the case of very large outputs, be finished at the manufacturers' works, with the stator core and windings complete; an important consideration from a constructional point of view. This is not the case with large output alternators of the low-speed type, owing to their great size and the limitations imposed by considerations of transport and shipment.

In marine work progress has been no less rapid, the direct coupled turbine of 1897 having been superseded in 1910 by the geared turbine, and the marine turbine has advanced in power from the 2000 shaft horse-power of the *Turbinia* to the 150,000 shaft horse-power of H.M.S. *Hood*.

The association of the turbine, essentially a high-speed engine with a low-speed propeller, put the marine turbine at a serious disadvantage so long as it remained directly coupled to the propeller shaft. The conditions for high propeller efficiency and high turbine efficiency were antagonistic, particularly in vessels of low speed. In spite of this difficulty, the direct coupled turbine soon established its superiority over the reciprocating engine for the propulsion of high-speed vessels, such as warships, liners, and channel steamers. A great advance was made by the introduction of gearing, the turbine having since been successfully applied to practically every class of vessel, and considerable improvement in economy effected by the increased efficiency of turbine and propeller when these are allowed each to run at its most efficient speed of rotation. The direct coupled marine turbine has now been completely displaced by the geared turbine.*

In the early direct turbines, the steam consumption was from 15 to 16 lb. per horse-power hour. With a geared installation of 3000 shaft horse-power consisting of three turbines coupled to the propeller shaft through double reduction gearing, the consumption has been reduced to less than 10 lb. per shaft horse-power hour for saturated steam, and 8 lb. per shaft horse-power hour with a superheat of 200° F.

Since the introduction of mechanical gearing the marine turbine has tended to become assimilated in type to the land turbine, the same conditions making for thermal efficiency, namely, high velocity ratios, high steam pressures and temperatures, and high vacua, being equally applicable. Designs have been prepared for several marine installations of various powers on these lines, and proposals made to certain ship-owning companies. As an example, it is estimated that by such extension of the thermodynamic cycle, and with geared turbines designed for the highest efficiency, the fuel consumption of a ship like the S.S. *Mauretania*, at present driven by direct coupled turbines, could be reduced by over 40 per cent.

* A full account of the adoption of geared turbines in the vessels of the British Navy was given by Engineer Comm. Tostevin, in a paper read before the Institution of Naval Architects in March, 1920.

Mechanical gearing has also supplied the solution to the difficulties of the continuous current turbo-dynamo. Moderate speed generators have now reached an output of 3000 kilowatts from a single machine, the current collected from a single commutator being as much as 6000 ampères, and, with the interposition of gearing, such generators can now be turbine driven.

Another considerable field of application for geared steam turbines is the driving of mills, such as textile, paper and jute mills. In these installations the high economy, powerful starting torque and smooth turning moment of the turbine are employed to great advantage. The last-mentioned property is particularly advantageous in the case of mills in which the power is transmitted to the different floors by means of ropes, from a main rope pulley coupled to the prime mover through gearing.

It is estimated that the output of all the steam turbines built, both land and marine, has now reached the great total of over 120,000,000 horsepower.

II. An analysis of the present position

It is now well understood that in steam turbine installations where steam economy is of prime importance, pressure compounding, or the placing of many simple turbines in series on the steam flow, is essential; further, that one of the fundamental requisites for the attainment of high steam economy is the operation of the turbine at an appropriate overall velocity ratio, or ratio of mean blade velocity to steam jet velocity.

It is known from the laws of thermodynamics that the thermal efficiency of a heat engine depends chiefly upon the temperature range; that, in accordance with Carnot's principle, the temperatures at which heat is supplied and withdrawn should be as widely separated as possible. A large temperature range must necessarily be accompanied by a large pressure range. This is exemplified in the case of steam by the fact that the latent heat, which is a large proportion of the total heat, is absorbed during the process of evaporation, and in order that this may take place at a high temperature, the boiler pressure must be high. Accordingly, the ratio of initial to final pressure in modern steam turbine practice is being increased, and in turbines under construction is in the neighbourhood of 1500 to 1. The corresponding temperature range is from about 470° F. saturation temperature, raised to 750° F. by superheating, down to about 75° F., or a maximum drop of about 675° F.

It is interesting to recall that a century ago, the temperature range of the best reciprocating steam engines was so small that the disappearance of heat by conversion into mechanical work passed almost unnoticed, and that neither James Watt nor Sadi Carnot appreciated its real significance, in spite of Count Rumford's famous experiments on frictional heat in 1798.

In modern turbine design, there arises the fundamental question of high steam velocities *versus* moderate or low steam velocities. From the point of view of reduction of the number of stages, and therefore the overall length of a turbine for a given capacity, it is clear that high steam velocities are all to the good, but in proportion as the steam velocity is increased, the essential benefit of pressure compounding is sacrificed, for the advantage of that principle does not rest solely on the fact that the hydraulic head under which each simple turbine of the series works is reduced, but also on the evidence that the efficiency of conversion of the available energy of the steam into mechanical work in the nozzles and blading decreases as the velocity is increased.

It will be recalled that De Laval's simple turbine, which to many would have seemed to offer the best solution of the economic utilisation of the enormous available energy in high-pressure steam, fell short of anticipated results owing to the heavy nozzle losses which occur when very high expansion ratios are dealt with in a single pressure drop.

The high economy shown by the published test results of the Ljungström radial flow double rotation turbine, which operates with higher steam velocities than are usual in parallel flow practice, has been attributed by some mainly to the reduction of steam leakage at the glands and balancing pistons by means of a multiplicity of very fine labyrinth packings.

Recent researches in the laboratory, with fixed nozzles, but without blade rings, have confirmed that nozzle efficiency diminishes with increasing steam velocity, and it remains to find out the causes of this loss of efficiency, and whether it may be possible to reduce or eliminate it.

Apart from the effect of jet velocity upon nozzle efficiency, there is also the fact that the stage efficiency of a steam turbine is lower with saturated steam than with superheated steam. Laboratory tests have proved that the weight of steam discharged by a nozzle using steam initially dry-saturated is appreciably greater than that which theory would indicate, and it is thought by some that the cause of decreased efficiency of stages using saturated steam is closely connected with this phenomenon. It is thought, in fact, that the steam, during expansion, becomes super-saturated, and does not give up so much of its latent heat as it would do under conditions of thermal equilibrium, with the result that the available heat for a given ratio of expansion is considerably less than that indicated by calculation on the basis of thermal equilibrium. This question is closely bound up with the thermodynamics of steam turbine design. All engineers engaged in such work are desirous of reducing the nozzle and blading calculations to as nearly an exact science as possible, and of placing test results of different turbines under different steam conditions upon a rational basis of comparison. Up to the present, although a good approximation has been attained, this problem cannot be said to have been completely solved, the influence of cumulative heat arising from expansion of the steam with a stage efficiency less than unity, and the above-mentioned fact of decreased stage efficiency in regions where wet steam is found having much complicated the problem. With regard to the design of blading of the impulse type, there are two schools of thought. One of these favours the proportioning of the passages so as to obtain as far as possible what is known as "free deviation"—in which the blades are arranged so as simply to deflect the steam jets issuing from the nozzles, without alteration to the steam pressure. The other favours the employment of a certain amount of pressure drop in the revolving blade passages. The latter procedure is conducive to higher stage efficiency, but involves some leakage radially over the shrouding, and also circumferentially in the case of partial admission; in such cases the same principle is used as has for many years been standard in the case of reaction turbines, namely the adoption of large radial clearances, but reduced axial shroud clearances.

The question has arisen as to whether a compromise between the so-called "Impulse" and the so-called "Reaction" principle can be made to give a higher efficiency than either principle alone, and it is interesting to note that impulse turbine builders and reaction turbine builders are approaching this question from opposite standpoints. At the present time reaction turbines are generally built with equal expansion ratios in the moving and guide rings of a reaction pair. Experiments have been made with a different division of the pressure drop, but up to the present, equal expansion ratios in the fixed and moving blading, and similar construction for these parts, have given the best all-round results.

Advance in the direction of increasing the thermodynamic value of the heat in the steam without unduly increasing the maximum temperature of the system is being made by increasing its mean temperature of heat reception. One method of effecting this is the adoption of regenerative feed water heating, by using steam tapped off from successive stages of the turbine. The necessity to heat the feed water, that is, to reheat the water of condensation, is the defect of the Rankine cycle. If, however, a perfectly regenerative process could be employed for heating the feed, the efficiency of the Rankine cycle for saturated steam would be brought up to that of the Carnot cycle. In such a process the steam tapped off has been expanded, doing work in the turbine, down to the temperature at which it is required to heat the feed water. Since a certain amount of heat is required for this purpose in any case, the work obtained from the tapped off steam is obtained merely at the expense of additional heat equal to the work done, or in other words at nearly 100 per cent. efficiency. Expressed in another way, the utilisation of some of the heat of the steam to heat the feed water reduces the amount of heat that is lost by condensation in the condenser.

A slight practical objection to such a system of feed water heating is the additional complication of the pipework involved, if a sufficient number of feed water heaters in cascade are employed to approximate to a thermo-dynamically reversible process; but even with two or three stages, very considerable heat economy can be effected. As a compromise, feed water heating by steam extraction from a few stages of the turbine, in conjunction with economisers, is generally accepted practice in modern power stations, but there is a tendency to develop this system further so as to attain higher feed water temperature by the steam heating. It then becomes necessary as an alternative or partial alternative to the economiser, to adopt some other means of absorbing the residual heat of the flue gases, such as a regenerative preheating of the air for the furnaces.*

As a second method of increasing the mean temperature of heat reception, an advance might be made if, after superheating to the maximum temperature, further heat were added so as to maintain isothermal expansion throughout the initial stages of the turbine. This could only profitably be carried to such a point that subsequent adiabatic expansion in the turbine to the condenser vacuum would leave the steam just saturated. Such a method (of preliminary isothermal expansion) was proposed many years ago, but the mechanical difficulties in the way of its complete realisation would appear insurmountable. An approximate solution, however, is being worked out at the present time; in some installations the steam, after a certain amount of expansion in the turbine, is extracted and raised again to a high temperature in a reheater before re-entering the turbine. Reheating has also the additional advantage of extending the superheated field, and therefore of diminishing the range wherein moisture exists with consequent loss by water resistance, supercooling, or increased viscosity. This process might, of course, be extended to include reheat at various stages of the expansion. With a single reheat a compromise must be accepted when selecting the stage at which the steam is taken from the turbine. If reheating is carried out too soon, then the heat added—although at a high mean temperature—is only of small amount. On the other hand, if reheating is carried out too late, then the heat added—although of greater amount—is received at a lower mean temperature, so that the mean temperature of reception of the cycle is not much improved.

* Feed heating in a single stage by partly expanded steam is a well-known procedure; it was proposed by James Weir in 1876, by Normand in 1889, and in progressive stages by Ferranti in 1906; with marine turbines it is general practice to utilise the exhaust steam from the auxiliary engines, at a pressure slightly above atmospheric for this purpose. Similar use can be made in power stations of any low grade heat that is available, such as in steam from turbine glands, ejector air pumps, etc.

If a curve be drawn showing the estimated percentage gain in thermal efficiency on a base of the proportion of expansion carried out before reheating, it is found that there is a fairly definite peak indicating the most appropriate reheating point. The maximum estimated gain in the case of a single reheat to the initial temperature of 750° F. is in the neighbourhood of 7 per cent.

Since reheating the steam increases the total available heat per lb. more blading is required in the turbine to maintain the same average velocity ratio; this and other considerations tend towards the desirability of carrying out the expansion in two or more separate cylinders, the exhaust steam from any one cylinder being reheated before entering the next. Where such reheating is adopted special means must be provided to prevent over-speeding of the machine, owing to the large volume of steam enclosed in the reheater and connecting pipes.

A third method of attacking the same problem is to increase the mean temperature of the heat reception by increasing the boiler pressure.

With a boiler pressure of 250 lb. per square inch, a final temperature of 750° F. after superheating, and a feed water temperature of 79° F., the mean temperature of heat reception is about 360° F. But if the boiler pressure is increased to say 2000 lb. per square inch, still superheating to 750° F., then the latent heat is received at the increased temperature of the boiler, and the mean temperature is considerably increased, namely to 480° F. If, in addition, we assume that the water is fed into the boiler at the boiling temperature, by means of feed heaters in cascade in each case, instead of at 79° F., the mean temperatures of heat reception are increased in the one case to 430° F., and in the other to 680° F.

Whilst such a pressure as 2000 lb. per square inch has not been actually adopted in any commercial plant, there seems to be no reason why it should not be practicable in a large steam turbine installation. In such a case there may be a small auxiliary high-pressure turbine, having several stages of small diameter, running at high revolutions, and after dealing with the first few expansions, exhausting into turbines working at ordinary pressures. Such high-pressure turbines are actually being built—one in this country for a pressure of 1500 lb. per square inch, and one in the United States of America for 1250 lb. per square inch.

With regard to the lower end of the temperature range, the temperature of the cooling water here imposes the limit. An exhaust vacuum equivalent to 29·25 inches of mercury (barometer 30 inches) can be realised under favourable circumstances. This, however, involves the necessity of providing ample blade area and passage way for the great volume of low-pressure steam.

Throughout the stages of a turbine, the criterion of efficiency is the velocity ratio, and the areas of the blade passages successively increase in such proportions as to provide the proper ratio between the blade velocity and the steam velocity. The axial component of the steam velocity is carried on from stage to stage without appreciable loss. It is even built up to higher values as the peripheral and steam velocities increase. Its final loss at the exhaust end of the turbine, however, is practically unavoidable, and with the high peripheral speeds adopted in most low-pressure blading at the present day, an exhaust blade area which gives a perfectly satisfactory velocity ratio may entail a loss of energy in the issuing steam, of considerable amount. Such loss is augmented if, for the sake of weight or cost, any sacrifice is made of the final velocity ratios, which sacrifice might otherwise be a perfectly justifiable compromise. It is evident, therefore, that the exhaust end of a turbine requires separate treatment in its design from the point of view of limitation of the leaving loss, that is, the energy lost in the issuing steam.

Broadly speaking, where high efficiency is the aim, this consideration imposes a limit upon the axial velocity of the steam in the final annulus. For a given vacuum, therefore, the area of the final annulus must increase in proportion to the output, and for a given output any increase in the vacuum involves an increase in the area of the final annulus. It will be seen, therefore, that any effort to extend the steam cycle by lowering the temperature of condensation brings at once into prominence the problem of the exhaust end.

Various expedients have been adopted for the purpose of surmounting this limitation. One method, commonly employed in both land and marine turbines, is to carry out the final stages of the expansion by divided flow, as for example in a double-ended low-pressure turbine, the steam entering at the middle and flowing in each direction towards an exhaust at each end. In such an arrangement two exhaust pipes are provided, and frequently two condensers. In another example of divided flow, some makers double only the last row or last few rows of blading, employing specially shaped passages to conduct the steam to the divided portion from the preceding stages.

The design of exhaust end advocated by Baumann is another attempt to overcome the same difficulty. In this device, instead of carrying over a part of the steam by special ports to additional blade rings, it is made to flow through inner annuli of the preceding blade rings, the vanes of these inner annuli being specially shaped to reduce the resistance involved to a minimum.

The area of the final annulus can be increased by increasing the blade height, and any means of extending the height permissible on a given diameter is valuable for this purpose. Thus in many turbines, the blades of the last stages are made of varying section and discharge angle, permitting a greater length ratio without unduly increasing the stress.

If no restrictions are imposed either upon the diameter of the exhaust end or upon the speed of revolution, a simple and probably the best solution of the exhaust area problem in large machines is to provide a separate low-speed turbine of large diameter for the final stages of the expansion, having blades of normal profile throughout, and correspondingly high efficiency. This low-speed turbine can be arranged in close proximity to the condenser, as illustrated diagrammatically in Fig. 1, so as to reduce to a minimum both the length and the resistance of the exhaust passages.

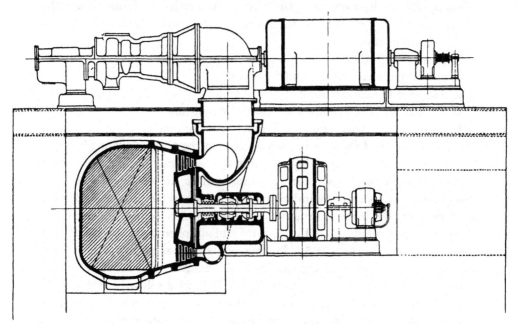

Fig. 1. Scheme for utilisation of high vacuum

In the exhaust passage itself, it is not sufficient to provide area to reduce the mean velocity of the steam passing through it to a moderate value, but the exhaust steam must be distributed into the passage uniformly, so that the velocity is everywhere nearly the same. Neglect of this requirement leads to unnecessary hydraulic losses, and may also cause troublesome blade vibration in the final stages.

Many attempts have been made to obtain the benefit of the diffuser principle, which has been so commonly adopted in the case of water turbines, but in the case of steam turbines without substantial success. In any attempt to carry this principle into effect, ample exhaust pipe area should be provided, with progressively larger cross section on the way to the condenser, so that the exhaust steam from the last row of revolving blades is gradually slowed down, with a possible recovery of static head.

The practical difficulties in the way of such recovery of pressure in the exhaust steam are very great, and the best that can be done in this direction would appear to be to utilise the leaving velocity of the exhaust steam to overcome the resistance to flow in the exhaust passages, so that the absolute pressure at entry to the condenser is the same as at the exit from the blading.

III. An application of the foregoing principles

Summarising the foregoing, there are four methods of improving the thermal efficiency of the thermodynamic cycle without increasing the maximum temperature:

1. Increased feed water temperature, with regenerative heating of the feed water by steam drawn from the turbine.
2. Increased boiler pressure.
3. Reheating of the steam after partial expansion.
4. Increased vacuum and means to utilise it effectively.

An application of these principles is made in the turbine plant of 50,000 kilowatts normal capacity to be erected in the new Crawford Avenue Power Station, Chicago, and now nearing completion at Newcastle-on-Tyne. In this plant there are three cylinders in series on the steam flow, each driving an alternator, the three alternators being electrically coupled in parallel.

Steam is generated at 600 lb. pressure and supplied to the stop valve of the high-pressure turbine at 550 lb. pressure, superheated to 750° F. After expansion in the high-pressure turbine to a pressure of 100 lb. above atmosphere, it is led back through a well-lagged pipe into a reheater in the boiler house from which it is returned to the intermediate-pressure turbine at a temperature of 700° F. The high-pressure and intermediate-pressure turbines run at 1800 revolutions per minute and drive alternators of 15,000 kilowatts and 30,000 kilowatts capacity respectively.

Further expansion takes place in the intermediate-pressure turbine to a pressure of about 2·0 lb. absolute, at which pressure the steam enters the low-pressure cylinder, to be expanded to the condenser vacuum of 29·25 inches of mercury (barometer 30 inches).*

The low-pressure turbine drives a 5000 kilowatt alternator at 720 revolutions per minute. Owing to the lower revolutions of this turbine, the area provided in the final stage is so large that blades of normal profile, but set at a slightly greater discharge angle, can be employed. The steam is condensed in twin surface condensers with the tubes vertical, and having

---

* The condensing water is obtained from Lake Michigan and its temperature is as low as 35° F. during the winter months.

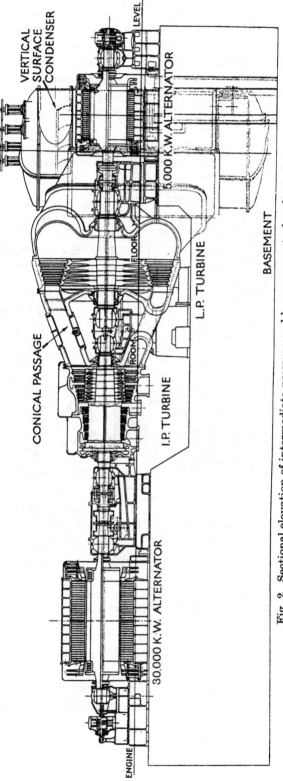

Fig. 2. Sectional elevation of intermediate-pressure and low-pressure turbo-alternators of a 50,000 kilowatt plant.

a total surface of 56,000 square feet. Fig. 2 is a sectional elevation of the intermediate-pressure and low-pressure units. The whole installation of turbines, alternators and condensers is illustrated in Fig. 3.

About 22 per cent. of the total steam entering the turbine is used for feed water heating; it is extracted from the turbines at three points, the feed water being heated up from 65° F. to 315° F. in three stages before entering the economisers.

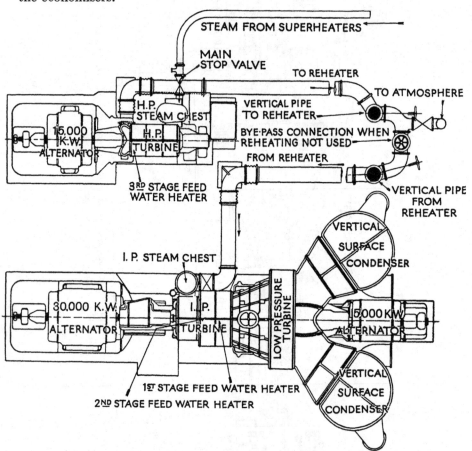

Fig. 3. Arrangement of 50,000 kilowatt turbo-alternator and
surface condensers, shown in plan.

A heat consumption of 10,265 British thermal units per kilowatt hour is anticipated, or a thermal efficiency (from steam to electricity) of 33·2 per cent.

For recovery of the heat in the flue gases, economisers are used for a final stage of feed water heating, and air preheaters, a regenerative process being employed in the latter to heat up the incoming furnace air.

The boiler plant is expected to have an efficiency of 86·5 per cent. After allowing about 3 per cent. for the total power absorbed by the auxiliaries, an overall thermal efficiency (from fuel to electricity) of 27·80 per cent. is anticipated. This is on the basis of a boiler pressure of 600 lb. per square inch. The thermal efficiencies, it is estimated, could be realised with higher boiler pressures, up to 2000 lb. per square inch are shown in the table.

Some details of the turbines and alternators will be of interest.

The three cylinders contain 64 simple turbines arranged in series, the first having blades $2\frac{1}{2}$ inches long and $\frac{3}{4}$ inch wide on a mean diameter of $30\frac{1}{2}$ inches, and the last having blades 40 inches long and $3\frac{3}{8}$ inches wide on a mean diameter of 160 inches.

In view of the low density of the steam, the passages from the intermediate-pressure turbine onward have been specially designed to have the minimum of resistance and to avoid any sudden enlargements which would produce eddies or shock. The intermediate-pressure turbine discharges its steam directly into the low-pressure turbine through a conical annular space surrounding the intermediate bearings, and the energy of the leaving velocity of the intermediate-pressure exhaust is thus conserved. (See Fig. 2.)

The maximum blade tip velocity is reached at the last ring of the intermediate-pressure turbine, where it is 760 feet per second; in the low-pressure turbine, which runs at 720 revolutions per minute, the maximum tip speed is 626 feet per second.

On the constructional side the most important feature of a turbine is its blading. There is probably no other part upon which so much ingenuity has been expended, and there are wide differences in practice both as regards materials used and the methods of manufacture and of attachment. The best known system is that in which the blades are made of rolled brass strip fixed into the rotor or casing by caulking spacing pieces between the blade roots. This system has been widely adopted in the past for both land and marine turbines, and has proved very satisfactory where the stresses and temperatures are moderate. When properly carried out this method gives a strength of attachment practically equal to the breaking strength of brass blades.

Many manufacturers and users now favour blades integral with their roots, and with such a formation higher stresses can be employed with suitable material. Such integral blades have until recently been formed by milling from the solid bar.

The roots are machined to suitable shapes or are serrated and attachment is effected by driving up the combined pieces in accurately machined or serrated grooves or slots.

Recent improvements in the methods of manufacture have, however, been developed at Heaton Works by means of which it is now possible to

Table showing the overall thermal efficiencies which it is estimated could be realised with increased boiler pressures, up to 2000 lb. per square inch.

(Based on the Callendar Tables and Formulae for the Properties of Steam.)

Column no.	1	2	3	4	5	6	7	8	9	10	11	12	13
Case no.	S.V.P. lb. gauge	S.V.T. °F.	°F. initial super-heat	Assumed reheat pressure lb. abs.	Reheat temp. °F.	Restored super-heat °F.	Exhaust vacuum in. Hg. Bar. 30·0 in.	Stage feed water heating from 65° F. up to (°F.)	Thermal efficiency (from steam to electricity)	Assumed boiler plant efficiency (including all aux.)	Overall thermal efficiency (from fuel to electricity)	Per cent. reduction in fuel consumption	Equivalent lb. of fuel oil per B.H.P. hour
I	250	750	344	65	700	402	29·25	360	31·6	84·0	26·52	—	0·501
II	500	750	281	100	700	372	29·25	420	33·5	83·5	27·97	5·45	0·475
III	1000	750	204	150	700	343	29·25	510	35·2	83·0	29·20	10·08	0·455
IV	1500	750	153	250	700	299	29·25	550	36·6	82·5	30·20	13·82	0·440
V	2000	750	114	400	700	254·5	29·25	600	37·5	82·0	30·75	15·90	0·432

S.V.P. = Stop valve pressure.
S.V.T. = Stop valve temperature.
Feed water heated to within about 50° F. of boiler temperature by means of steam withdrawn from turbine blading at several points.
Boilers equipped with economisers or air pre-heaters or both, so as to recover heat from the flue gases and maintain high thermal efficiency.
Dynamo efficiency taken as 96·5 per cent. at full rated output.
Gross calorific value of fuel oil taken as 18,500 B.Th.U. per lb.

PLATE VI

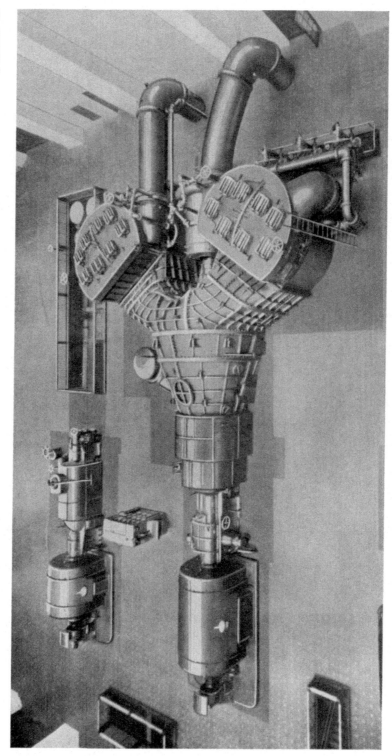

Parsons 50,000 kW Turbo-Alternator in Chicago Power Station

roll the blade integral with its root or spacing piece, and at the same time to form a radius at the junction of the two. This method of manufacture has the advantage of retaining the valuable feature of working the material so that the grain or fibre is developed in the right direction for strength. It also subjects it to a strenuous test and has been found to detect faulty material with considerable certainty.

For these reasons such methods of rolling give a superior product to that attainable by milling from the solid or by processes of drop forging or stamping, and are at the same time less costly.

In the 50,000 kilowatt turbine previously mentioned, the blades throughout are of mild steel, formed integral with their roots by this method of rolling. They are fitted in serrated grooves in both rotor and cylinder, being inserted and driven up tight. The blades of the high-pressure turbine, and of the first half of the intermediate-pressure turbine, are fitted with "end-tightened" shrouding made of manganese copper and brazed over tenons on the ends of the blades. "End-tightened" blading was introduced commercially in the year 1912, and has been widely adopted in land turbines of the reaction type and in several mercantile marine installations. In this type of blading the necessary running clearances are maintained in an axial direction, instead of radially, so that they can be made small, and adjusted and regulated under the control of the thrust bearing, which determines the position of the turbine shaft axially in its cylinder. The minimum radial clearances of the portion of the blading that is end-tightened is $\frac{1}{4}$ inch, whilst the radial clearances in the remainder of the blading range from $\frac{1}{8}$ inch up to $\frac{1}{4}$ inch.

Each turbine shaft is fitted with an emergency governor and these are interconnected so that any one of them will trip all steam admission valves simultaneously and at the same time open a large vacuum breaking valve. In order to meet the special conditions that arise when reheating is adopted, the governing of the turbine has been arranged so as to avoid any danger of over-speeding due to stored steam. This is done by the provision of an independent governor valve on the inlet to the intermediate-pressure turbine, which receives steam from the reheater. The governor controlling this valve is set to operate at a speed about 2 per cent. in excess of that of the main governor, and with this lag will control the rate of steam admission to the intermediate-pressure turbine whenever the load on the plant is suddenly reduced. In case this valve should close for any reason, whilst the main high-pressure inlet remained open, a relief valve of ample area is provided on the exhaust pipe of the high-pressure turbine and set to open at 130 lb. per square inch above atmosphere, whilst an increase in pressure to 140 lb. will operate an additional tripping device and shut off the main steam supply.

The alternators are wound for a three-phase supply at 13,000 volts, and embody the latest British practice. No main field regulators are used in their operation, the regulation being carried out by means of well-designed exciter field rheostats.

The ventilation is effected by means of the closed circuit system employing surface air coolers and separately driven fans, which have a much higher efficiency than high-speed fans integral with the rotor shaft.

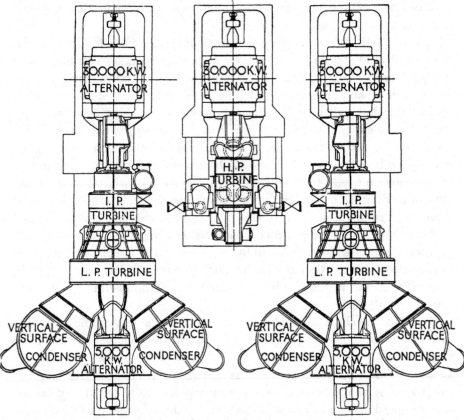

Fig. 4. Plan of scheme for increasing the capacity of the turbo-alternator in Fig. 3 to 100,000 kilowatts.

The stator frames are being built in one piece, and the alternators are to be shipped completely wound.

Since, when starting up the plant, there will be insufficient steam to drive the low-pressure turbine, the alternator coupled to this turbine will be run up as a synchronous induction motor. The three alternators will then be synchronised together and operated as one unit, giving a combined normal output of 50,000 kilowatts.

The Crawford Avenue Power Station has been laid out for a total capacity of 500,000 kilowatts. Fig. 4 shows how readily such an installation as the one described can be developed into one of 100,000 kilowatts without increase of the largest component units by replacing the high-pressure unit by one of double the steam capacity, and by duplicating intermediate-pressure and low-pressure units. There would thus be three alternators of 30,000 kilowatts output and identical design, and two smaller alternators of 5000 kilowatts driven by the low-pressure turbines.

SOME INVESTIGATIONS INTO THE CAUSE OF EROSION OF THE TUBES OF SURFACE CONDENSERS

Institution of Naval Architects, April 6, 1927

Before reading this paper it may be explained that only a few months ago it was decided to commence some experiments on the possibility of the erosion of condenser tubes by the action of cavitation and water hammer. The time has been short, and in some respects the experiments are incomplete, but it is hoped that as far as they go they may prove helpful in the consideration of the causes of premature failure of condenser tubes.

The problem of the failure of the tubes of surface condensers is so well known that it seems only necessary here to allude to a few of the salient facts bearing upon the question. Whilst no adequate explanation has as yet been assigned for such failure, many attempts have been made to overcome it. On the theory that it is caused by electrolytic corrosion, counter-electromotive force has been applied to neutralise the action; on the theory that it arises from chemical action, such means as coating the inside of the tubes with a bituminous paint or with a protective scale of oxide have been tried. None of these measures have proved to be more than palliative in their effects, and sometimes not even that. In a few cases more expensive metals are being adopted for the tubes, such as cupro-nickel or monel metal. For many years bell-shaped mouths have been fitted to the inlet ends of condenser tubes to produce a smooth stream-line flow into the tubes,* and this is claimed to prevent the formation of a *vena contracta* by the sharp square entrance to the tube and the liberation of oxygen and other occluded gases from the water which cause corrosive action at the tube entrance.†

In most cases of failure it has been observed that certain of the tubes are eroded, corroded, pitted, and eventually holed, but always or nearly always from the water side, that is to say, from the inside surface. The action is usually confined to tubes in certain regions of the condenser tube plate, and when such tubes have been replaced from the common batch they have usually failed just as quickly, whilst in other positions in the tube plate the original tubes have remained immune, so that there appear to be distinctive areas in which the conditions favour corrosion or erosion. Further, it appears that the action is greatest at the entry ends of the tubes, frequently extending

---

* *Vide* P. C. Parker's remarks in the discussion on Dr Bengough's paper, *Transactions of the North-East Coast Institution of Engineers and Shipbuilders*, Vol. XI, Part 2, 1923.

† *Vide* Serial Report of the Prime Movers' Committee, 1926–7, of the National Electric Light Association of New York, U.S.A.

only a few feet from the mouth. These facts seem to suggest that the cause may be hydrodynamical rather than electrical or chemical, and in the light of the experiments described in this paper they seem to indicate that these areas of erosion or corrosion are associated with regions of turbulent motion of the water in the water box.

The object of the present paper is to give an account of some experimental investigations which have been designed to throw light upon the actual conditions attending such regional disturbances in the water box, and the manner in which they affect the character of the flow through the tubes in their vicinity, also to investigate the possibility that the pitting of condenser tubes may in reality be due to water hammer of collapsing vortices, which is known to be a potent cause of erosion of screw propellers and of the impellers of centrifugal pumps and water turbines.

A paper was read before the Institution in 1919 by the present author and Mr Stanley S. Cook on the "Cause of Erosion of Propellers". The paper recounted the investigations carried out by the Propeller Erosion or Corrosion Committee of the Board of Invention and Research. In that paper experiments with a water-hammer cone were described, and it was shown that by concentrated collapse of a vacuous cavity in water under atmospheric pressure small holes could be punched through brass discs, demonstrating the production by this means of shock pressures as high as 150 tons per square inch. The conclusions of the Committee were that the powerful and erratic erosion experienced with propellers was due to vacuous cavities in the water arising from eddies caused by struts or bossing, or by excessive slip of the propeller itself, which cavities collapsing on the surface of the propeller blade caused repeated blows and erosion of the metal.

The first of the present experiments was designed so that the behaviour of the water in the water box and also in the tubes could be observed and, if desired, photographed.

An apparatus consisting of a small wooden water box, the front being of plate glass, with a brass tube plate for ten tubes was constructed. To facilitate observation the first 3 feet of the four tubes of the central vertical row were made of glass also, the tubes of the side rows were omitted, and instead retarding plugs were fitted giving the same discharge as a 15 feet length of condenser tube. The discharge from these side holes was led into a separate chamber immediately behind the tube plate. The four central glass tubes passed through the back of this chamber, and were joined at their ends by clamped rubber-tube couplings to brass tubes 12 feet long, making a total tube length of 15 feet. They were also surrounded by a rectangular water tank with plate glass sides so that (the refractive index of glass and water being nearly the same) the water inside the tube could be seen without appreciable refraction. The discharge from the long tubes was taken into a measuring tank with a glass front, so that the trajectory of the water issuing from the tubes could be observed.

Closely fitting between the front plate glass of the water box and the tube plate, an inner and movable water box of circular shape was fitted into which the water was led tangentially through a socket fitting. This inner water box was moved by a lever passed through a stuffing box, so that it could be moved up and down across the face of the tube plate. An air vessel on the water supply to the box maintained a steady pressure of about 10 lb. per square inch. With water entering the water box at a speed of 8 feet per second, which was also the speed of the water through the tubes, a vortex filament was observable near the centre of the circular inner box, extending across from the glass plate to the tube plate. By moving the box this vortex could be brought over the mouths of tubes 2 and 3 of the four central tubes. The tubes 1 and 4, top and bottom, being only presented to the outer zones of the vortex filament.

The following results were observed when a vortex filament was in this way brought over the mouth of a tube, through which up to that moment the water had been flowing with a speed of 8 feet per second.

1. The vortex was drawn into the mouth of the tube.

2. The flow in that tube was partially arrested, the discharge being reduced in a fraction of a second to about one-fifth of its previous quantity.

3. The vacuum at the core of the vortex brought over the tube mouth was momentarily intensified by the arrest of momentum of the water already in the tube.

4. A vacuous thread extended momentarily into the tube for about half the length of the glass portion, and then collapsed irregularly at various points.

5. Silvery looking fringes appeared momentarily on the entry edges of all the four tubes whether under or near the vortex.

6. It was observed that at the instant when the vortex was brought over the tubes 2 and 3 it developed a momentary bifurcation into the neighbouring tubes 1 and 4; the bifurcation being of such short duration as to be difficult to see.

The apparatus was next modified for continuous working in order to ascertain if pitting of the tubes could actually be produced in this manner. The glass tubes were replaced by tubes of brass, continuous for the first 12 feet of their length, and then extended to 15 feet by attached 3 feet lengths of similar brass tube. A solenoid, with a make-and-break relay, was fitted to operate the lever to oscillate the inner water box, at a rate of about 15 cycles per minute, and a motor-driven centrifugal pump and drain tank maintained a continuous circulation, the system being filled with brackish water from the river Tyne.

The solenoid-operated lever was so adjusted by guides and stops as to bring the vortex filament alternately over the mouths of the second and third of the four central tubes, so that a comparison of these tubes afterwards with the first and fourth, which were not subjected to the action of the central part of the vortex filament, would enable any difference in effect to be definitely attributed to this action. The apparatus was run for 1450 hours, or about two months' continuous work, after which the four tubes were taken out and the first 3 feet of each sawn in two for examination. Pitting was observed at the entry of all the four tubes, extending from about ⅛ inch to a few inches along the several tubes, beyond this the tubes were apparently unaffected. In one of the tubes several pits penetrated about half-way through the thickness of the tube. There appeared to be very little difference in the amount of pitting in the two tubes subjected to the vortex filament, and those subjected only to the rapid transverse flow of water across their mouths, and this will be referred to later.

The experiments will, however, be continued with varied conditions, including transfer of heat across the tubes to the water as in ordinary condenser practice, and it is hoped that these may throw further light upon the subject, and produce experimentally the erosion of the tubes in the central portions of their length.

In order to investigate the character of the regional disturbances occurring in a water box of usual type a further experiment was made. A model water box was constructed, about a quarter full size of an actual condenser box, with a tube plate for one hundred tubes of ⅝ inch diameter. These tubes were bent and spread out in groups so that the discharge from each tube could be observed. The water box was fitted with a glass front. Town water was used, and was circulated by a centrifugal pump. The velocity in the tubes was about the same as in an ordinary condenser, i.e. about 8 feet per second. The tubes being of different lengths retarders were fitted near their mouths to give a discharge equivalent to tubes of 15 feet length, and to maintain the standard velocity in all the tubes. Regions of strong turbulence were distinctly observed in the water box, and the vortex centres were seen to move about rapidly from tube to tube over certain areas. The discharge from some of the tubes was found to show sudden variation; this variation corresponded, as regards the tube positions, with the vortex motion in the box, although the sudden drop of velocity was only about one-half that observed in the previous experiments with the circular water box. It, however, must be borne in mind that the vortices, in relation to the size of tubes, would only be of about one-quarter the size they would be in an average full-sized condenser, and also smaller than the single vortex in the circular water-box experiments.

In conclusion, it is suggested that the experiments, as far as they have been carried out, seem to lead towards the view that one of the chief causes of the erosion of condenser tubes may lie in the turbulent motion of the water in the water box of the condenser, and that this cause could be mitigated by enlarging the area of the inlet passages so as to reduce the velocity of the water as much as possible before entering the box, or be entirely removed by arranging in the water box a grid of suitable mesh and with passages of depth somewhat greater than their width. By the passage of the water through this grid such turbulence is practically destroyed. Grids of this kind have been tried in conjunction with the experimental apparatus and it is found that with such a grid, when the whirl chamber is moved across the tube plate exactly as before, the velocity at the discharge end of the tubes remains perfectly steady, and no vortex motion is visible in the box either behind or in front of the grid, also in the model water box, the velocity of discharge from the several banks of tubes is rendered practically constant.

It may be desirable to add that though the experiments have not as yet proved as conclusive on all points as was expected when deciding to read this paper, yet it now seems clear that with modifications of the apparatus they can probably be made so; for example, the last experiment has revealed the turbulence in the water box and the rapid sweep of the vortices over the mouths of the tubes, and if the velocity of their sweep is compared with the velocity of transit of the circular water box, moved by the solenoid, it is evident that the latter has been moving far too slowly to cause proper cavitation in the tubes 2 and 3 in the second experiment. To confirm this (and since this paper was written) an experiment has been made of replacing the vortex by a small valve with a leak through it so as partially to block the mouth of a 15 feet condenser tube, and by varying the amount of the blocking, and the rapidity of closing this valve, it has been shown, when part of the condenser tube is replaced by a glass tube, that the position of the water hammer along the tube can be regulated.

RECENT PROGRESS IN STEAM TURBINE PLANT

Read before the British Association, S. Africa, Section G, on July 31st, 1929

INCREASED FUEL ECONOMY AND OUTPUT PER UNIT

In the development of modern steam turbines, increased fuel economy is being realised by raising the initial steam pressure and temperature, by regenerative heating of the feed water so as to reduce the percentage of heat rejected in the condenser water, and by the use of air preheaters and economisers to extract the residual heat of the flue gases. Reheating of steam after partial expansion is also being tried as a means of increasing the economy; this is dealt with separately below. Improvements in turbine efficiency have been continuously made as the output per unit has increased. Above 5000 kilowatts the effect of size on efficiency is less marked. The overall heat economy of a large plant equipped with every modern auxiliary device will, however, generally be higher than that of a small plant.

These developments, in conjunction with the demand for units of larger and larger output, have emphasised the value of a more economical utilisation of the materials of construction of such turbines. The limiting factor to the output of a condensing steam turbine of high efficiency is well known to be the area of the blade annulus of the final blade ring at the exhaust end that is permissible for a given speed of revolution. By the use of higher quality steel for the discs, stronger blades and blade fixing and improved mechanical design, this permissible annulus has been increased about ninefold, and the output in rather greater ratio.

With a triple exhaust an output of 40,000 kilowatts at 3000 r.p.m. can now be obtained, and this is about the present-day limit of alternator capacity at that speed. At 1500 r.p.m. an output of 40,000 kilowatts can now be obtained from a single flow turbine, and, by double ending the low-pressure blading, 80,000 kilowatts can be obtained from a single machine.

Where the feed water is heated by steam tapped off from the turbine, the output obtainable from a single turbine is still further increased by the amount of the additional power derived from the partial expansion of the tapped off steam before extraction. In this way turbines can now be built to have an output of 100,000 kilowatts at 1500 revolutions with a double-ended low-pressure stage, and 150,000 kilowatts with triple low-pressure stages in a tandem cylinder turbine.

REHEATING OF STEAM AFTER PARTIAL EXPANSION

Extracting steam from the turbine at a suitable point, after partial expansion, and passing it through a reheater, can be shown on theoretical grounds to give some increase in thermal efficiency. This method has been tried on a large scale both in England and the United States of America. Since, however, most large modern turbines use cascade feed-water heating by steam extracted from the turbine at several stages of the expansion, when reheating is also adopted it becomes a matter of considerable difficulty in a large base load plant to obtain directly comparative test data. Such reheating was adopted in the 50,000 kilowatt reaction turbine and alternator erected in the Crawford Avenue Power Station of the Commonwealth Edison Company at Chicago in May, 1925, which turbine and alternator were designed and built at Newcastle-upon-Tyne.

Reheating to 700° F. by means of flue gases is now being adopted at the Barking Power Station of the County of London Electric Supply Company, Limited.

Reheating by means of live steam at boiler pressure instead of by flue gases has also been tried. The general impression appears to be that this method of reheating at such temperatures as have been tried is ineffective. If, however, reheating can be carried out to the initial temperature, or, say, to 750° F., by heat exchangers in the flue gases, then the benefit from the point of view of fuel economy is appreciable. On the other hand, there are the disadvantages of the complication of pipes and valves, and the necessity for overspeed safety devices on account of the large steam capacity of the reheater system.

IMPROVEMENTS IN SURFACE CONDENSER DESIGN

Modern improvements in surface condensing plant are in the direction of greater attention to the distribution of the exhaust steam among the tubes, with the object of reducing to a minimum the difference between the pressure in the exhaust and the lowest pressure in the condenser, yet retaining the temperature of the condensate at as nearly as possible that of the exhaust steam and promoting de-aeration of the condensate. According to one method, suitably situated channels and baffles are provided between the tubes by which means the exhaust steam is given more direct access to the condensate at the bottom of the condenser, and air extraction is shielded from the steam by banks of cold tubes. Another method is to pass the feed water through widely spaced tubes in the exhaust passage above the condenser.

PROGRESS IN HIGH-SPEED ALTERNATOR CONSTRUCTION

The maximum output of high-speed alternators at a given rate of revolution has increased nearly fourfold during the last decade; for instance, at 3000 r.p.m., from about 10,000 kilowatts to 40,000 kilowatts. This advance has been mainly secured by lengthening the rotor body and increasing its diameter, the latter being made possible by the use of higher tensile steel in the end caps.

Whereas, in steam turbine construction, it is possible to meet limitations as to length and diameter of rotor by adopting a two-, or even three-cylinder design, and duplicating the low-pressure stages, a similar change in alternator design is not practicable. For outputs of over 40,000 kilowatts and a frequency of 50 periods, alternators are now therefore designed for a speed of 1500 r.p.m., and the largest output at present obtainable at this speed is about 150,000 kilowatts, thus coinciding with the maximum obtainable from a steam turbine at the same speed.

In order to obtain a satisfactory design of alternator of this size and speed, the voltage should be increased beyond the value which, up till the present, has usually been adopted. Electrical engineers have been aware of the advantages of increased voltage and the tendency in this direction can be readily observed in the practice of the last thirty years. The highest generator voltage in the early alternators was from 2000 to 5000 volts, gradually increased to 11,000 volts by 1905, and subsequently to 14,000 volts. In South Africa, the same tendency is apparent. Alternators built a few years ago for the Electricity Supply Commission, for their Witbank Power Station, were designed for 6600 volts, whereas in the Salt River Power Station, at Cape Town, electrical energy is being generated at 11,000 volts. The Victoria Falls and Transvaal Power Company are also increasing their generating voltage to 11,000.

The need for higher voltages has led to considerable research work, and as a result a 25,000-kilowatt turbo-alternator has been built, generating at 33,000 volts. It has been in commercial operation since last August, in the new power station of the North Metropolitan Electric Supply Company, Limited, at Brimsdown, North London. A special feature of this design is the adoption of a concentric type of stator conductor containing three helically stranded copper conductors separated by mica insulation and joined up in series.*

By the method of winding, the highest potential is reached in the innermost of the three conductors, but the potential difference between this

* "Direct Generation of Alternating Currents at High Voltages." Paper read before the Institution of Electrical Engineers, March 21, 1929; see also *The Engineer*, Vol. CXLVII, No. 3821, p. 383 and *Electrical Times*, Vol. LXXV, No. 1953, p. 475.

and the next conductor is only about one-third of the total potential to earth, and the same applies to that existing between the inner and outer conductors, and between the outer conductor and earth. By the use of this type of conductor, it is possible to obtain the increased phase voltage without increasing the voltage gradient across the insulation, so that the electrical stresses in the insulation are no greater than those which, for a number of years, have proved to be perfectly safe. A section through one such conductor bar is illustrated in Fig. 1. It resembles an ordinary concentric cable, except that the insulation between the concentric conductors is micanite. Apart from the stator winding and mechanical details which have been modified to meet the special features in the design, this 25,000-kilowatt alternator is of standard construction. Since it was installed, it has operated continuously up to its maximum load at voltages

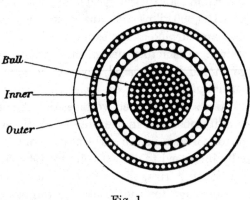

Bull

Inner

Outer

Fig. 1

between 34,000 and 35,000, and has withstood the most severe faults of the large overhead and underground network to which it is coupled without any sign of distress. The control and regulation have proved in every way satisfactory.

The above example is a pioneer step in direct generation at high voltages. So far only the fringe of possibilities of such a design has been touched; it would seem that in the natural course of progress, a reliable high-voltage alternator will become an essential factor in the rapid increase in size of power stations and their interconnections.

The advantages of high-voltage generation are not confined to the alternator. With the consequent smaller current, the number of cables which lead the current from the alternator are reduced, giving considerable economy of space for accommodating the cables and sealing ends mounted below the alternator terminals. The tunnel in which the cables are run through the foundation block is smaller, and its construction is simplified. There is also a reduction in the cost of switchgear, which for the transmission of large powers increases rapidly with the current. Further, in a large local network of high-voltage transmission to which the alternator can be directly coupled, there is the elimination, or partial elimination, of the usual step-up transformer plant, and the space and building required to accommodate them.

While only alternators have so far been dealt with, other units the size of which is continually increasing, such as motors, motor generators, synchronous condensers, etc., may be economically designed for high voltages and coupled direct to the network without the use of transformers.

Large alternators of the present day are ventilated by separately driven fans. It is interesting to note that one of the first plants fitted with this system of ventilation was a 1500-kilowatt alternator built for the East Rand Proprietary Mines. The problem of alternator cooling is of greater difficulty for many plants in South Africa on account of the high altitudes of the stations, the density of air at 6000 feet up for instance being about 20 per cent. less than at sea level.

IMPROVED STEEL INGOTS FOR LARGE FORGINGS

A subject of considerable importance in turbo-alternator construction is the strength and reliability of large steel forgings for turbine shafts and alternator rotors. The standard shape of ingot normally employed for casting molten steel is one in which the height is considerably greater than the breadth or diameter, with the larger cross-section at the top. In making ingots of this nature, a cast-iron mould is used on which is imposed a "head" lined with refractory material. As during cooling the crystals grow from the bottom and from the side walls, the solidifying metal contracts, and the still fluid metal in the "head" flows down and fills up the space formed by the contraction. During the course of the solidification, however, sulphide and non-metallic inclusions are thrown out of solution, and are forced towards the centre of the ingot. This segregated material, instead of rising to the surface and being rejected, becomes trapped in the freezing crystals, forming lines of weakness in the metal. A sulphur print taken from a large ingot cast in a mould of the ordinary type usually shows a line of segregates of inverted V-type approximately half-way out from the central axis, and numerous small V-shaped segregates down the central axis.

A new method has now been devised* for the purpose of making improved steel ingots, in which the molten metal is poured into a mould, having its horizontal dimensions greater than its vertical. Thick refractory material covers the sides of the mould, and there is a bottom chill of large

---

* "A New Method for the Production of Sound Steel." Paper read before the Iron and Steel Institute, May 3, 1929.

dimensions. The mould is pre-heated to a temperature approaching that of molten steel before the pour is made, and heat is continuously supplied to the upper surface of the molten metal after pouring, ensuring that the upper layers are the last to solidify. Fig. 2 shows the arrangement of the mould with a refractory cover and the openings for the oil burners. Cutting an ingot through the central vertical section and examining the macro-structure, after etching with Humfrey's reagent, brings clearly into evidence the existence or non-existence of segregates and lines of weakness of the structure. Fig. 3 is from an ingot cast by the new method just referred to. It shows the entire absence of such lines in the latter. The uniformity of structure is very marked.

With the ordinary method, as is well known, the larger the ingot the greater is the segregation and axial weakness, whereas in the new method there is no axial weakness, and an almost complete absence of segregation, however large the ingot. The only limitation to the size of solid forging is the equipment at the forge for handling very large ingots. It will be seen that this method provides a means of producing sound ingots for the manufacture of the largest turbine shafts and alternator rotors, and with the reliable forgings thus available there is actually greater safety to-day in large turbo-alternator forgings than there was in similar parts for plants of moderate output manufactured a few years ago.

MARINE PROPULSION

In marine propulsion with steam turbines, since the introduction of mechanical gearing in 1910 for transmission of large powers from high-speed turbines to low-speed propellers, thus giving independence of design, and enabling both propellers and turbines to be designed for their highest efficiency, there has been continuous progress in economy of fuel consumption.

The arrangement of turbine most favoured in recent practice is one which consists of three cylinders in series with three separate pinions gearing into a common wheel on the propeller shaft. This disposition of the turbines is specially appropriate in conjunction with the improved thermodynamic systems already referred to in connection with land practice, utilising high boiler pressure and temperature. It allows the high-pressure and high-temperature portion of the expansion to be confined to a small high-speed turbine. It permits of easy provision for regenerative feed heating, gives turbines which are readily accessible for overhauling, and economises in the dimensions of the gearing, because, with the increased number of pinions, the power that has to be transmitted by any individual pinion is reduced.

The first example of an arrangement of this type was that fitted in

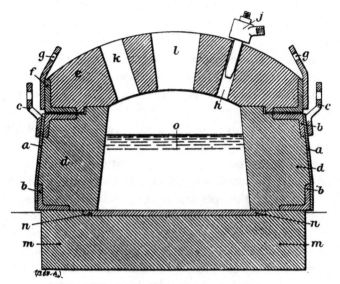

Fig. 2. Section of ingot mould

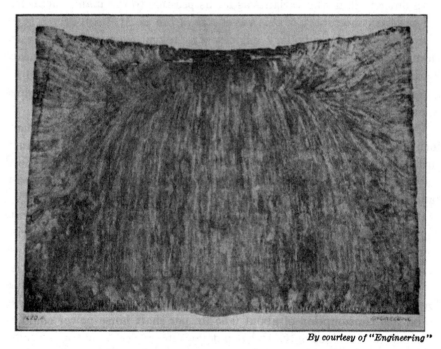

By courtesy of "Engineering"

Fig. 3. Etched macro-section of 20-ton ingot

the *Orama* by Messrs Vickers in 1924, a twin-screw vessel of 19,000 shaft horse-power at 19¾ knots. Each set of turbines had to develop 9500 shaft horse-power.

In order to demonstrate that a pressure of 550 lb. per square inch was not too high for marine work, an installation of turbines using steam at this pressure was fitted in the pleasure steamer *King George V* in 1926. This vessel has since been running daily on the Clyde during two summer seasons, and has shown an economy in coal consumption of 30 per cent. over similar vessels on the same service.

The experiment showed that such installations present no difficulties that cannot be met by suitable provision. In this pioneer installation, the boiler pressure was 550 lb. per square inch, the steam was superheated to nearly 800° F., and the turbine developed 3500 shaft horse-power. It will be clear that the economic effect of high pressure and temperature would have been considerably greater in an installation of larger power. The experiment on this vessel, however, was undertaken not only in order to obtain a higher efficiency, but to explore the practical application of high temperature and water-tube boilers to marine work and to ascertain what practical difficulties, if any, had to be overcome in this connection.

In order to limit the variants as far as possible to the main machinery, the auxiliary machinery of this vessel was of the usual steam-driven type. The steam for the auxiliary engines was taken from the superheater through an internal pipe in the saturated steam drum, by which means its temperature was reduced without loss of heat to the system, and all danger to the superheater, through overheating when the main engines were shut down, was thus avoided.

The auxiliary engines exhausted at a back pressure of 5 lb. per square inch above atmospheric, and their exhaust steam was condensed in a surface feed water heater, thereby raising the temperature of the feed water up to 200° F. An additional high-pressure heater utilising steam tapped from the turbines further raised the feed temperature to 300° F.

In proposals which were put forward by the Parsons Marine Steam Turbine Company in 1925 for the adoption of high pressures for marine work, with water-tube boilers with boiler pressure of 500 lb. per square inch, maximum temperature 700° F., feed heating to 350° F. and effective air preheating to give a boiler efficiency of about 84 per cent., it was forecast that, with a set of turbines developing 5000 shaft horse-power, an oil fuel consumption of about 0·55 lb. per shaft horse-power could be obtained for turbines only, and about 0·65 lb. per shaft horse-power for all purposes including auxiliary machinery. These estimates have been substantiated by the results obtained in some recent steamers built for the Canadian Pacific Railway Company, an account of which is given in a paper read by Mr J. Johnson before the Institution of Naval Architects in the spring of this year. The Table gives figures which are extracted from that paper.*

Fuel consumption of modern marine turbine installations

Ship	Shaft horse-power on service	Daily con-sump-tion of oil	Fuel per shaft horse-power hour	Pressure at engines	Tem-perature at engines	Vacuum
		tons			° F.	ins.
Empress of Australia	20,346	142	0·65	190	580	29
Duchess of Bedford	16,000	97	0·57	340	670	29
Duchess of Atholl	16,100	99	0·58	340	680	29

The fuel consumption given here includes that expended in working the main engines and the whole of the auxiliary machinery (including steering gear) incidental to the propulsion of the ship.

The development in marine work of higher pressures and temperatures is indicated by the fact that there are now built, or building, mercantile installations of about 500,000 total horse-power, with boiler pressures ranging from 350 to 550 lb. and steam temperatures of 700° F., and war vessels of over 650,000 horse-power, with pressures ranging from 325 lb. to 500 lb. and steam temperatures from 600° to 750° F.

It is reasonable to look forward to still further improvements in overall fuel economy in marine steam propulsion by developments in the same direction, and by the employment of mechanical stokers or the use of coal in a pulverised condition.

* "The Propulsion of Ships by Modern Steam Machinery."

PART II

EXPERIMENTS ON CARBON AT HIGH TEMPERATURES AND UNDER GREAT PRESSURES, AND IN CONTACT WITH OTHER SUBSTANCES

From the Philosophical Magazine *for September,* 1893*

The primary object of these experiments was to obtain a dense form of carbon which should be more durable than the ordinary carbon when used in arc lamps, and at the same time to obtain a material better suited for the formation of the burners of incandescent lamps.

There were a considerable number of experiments made in which the conditions were somewhat alike, and many were almost repetitions with slightly varying pressures and temperatures. They may, however, be divided into two distinct classes: the first, in which a carbon rod surrounded by a fluid under great pressure is electrically heated by passing a large current through it; the second, in which the liquid is replaced by various substances such as alumina, silica, lime, etc.

The arrangement of the experiment was as follows: A massive cylindrical steel mould of about 3 inches internal diameter and 6 inches high was placed under an hydraulic press; the bottom of the mould was closed by a spigot and asbestos-rubber packing—similar to the gas check in guns; the top was closed by a plunger similarly packed; this packing was perfectly tight at all pressures. In the spigot was a centrally bored hole into which the bottom end of the carbon rod to be treated fitted; the top end of the carbon rod was connected electrically to the mould by a copper cap which also helped to support the carbon rod in a central position. The bottom block and spigot were insulated electrically from the mould by asbestos, and the leading wires from the dynamo being connected to the block and mould respectively, the current passed along the carbon rod in the interior of the mould.

The fluid was run in so as to cover the rod completely. The plunger was then free to exert its pressure on the liquid without injuring the carbon. The pressure in the mould was indicated by the gauge on the press.

EXPERIMENTS. CLASS I

Among the liquids tested were benzene, paraffin, treacle, chloride and bisulphide of carbon.

The pressures in the mould during the several experiments were main-

* From the *Proceedings of the Royal Society*, Vol. XLIV. Communicated, with an additional note on Diamond Manufacture, by the Author.

tained at from 5 to 15 tons per square inch; the initial size of the rod was in all cases ¼ inch, and the current from 100 to 300 amperes.

Results. In some of these experiments a considerable quantity of gas was generated, and the press had to be slightly slacked back during the experiment to accommodate it and maintain the pressure constant.

In all cases there was a soft friable black deposit of considerable thickness on the carbon.

In no case was the specific gravity of the carbon rod increased by this process. There was no change in appearance of the fracture, excepting when chloride of carbon had been the fluid; it was greyer in this case.

The rate of burning of samples placed in arc lamps was not diminished by the process. Various rates of deposition were tried, but with the same result; and the conclusion seems to be that under very high pressures, such as from 5 to 15 tons per square inch, the deposit of carbon by heat from hydro-carbons, chloride of carbon, bisulphide of carbon, treacle, etc., is of a sooty nature, and unlike the hard steel-grey deposit from the same liquids or their vapours at atmospheric or lower pressures.

Experiments. Class II

In these experiments the asbestos-rubber packing was omitted, the plunger and spigot being an easy fit in the mould. A layer of coke powder under the plunger formed the top electrical connection with the rod.

No. 1. Silver sand or silica was run around the carbon rod, and pressures of from 5 to 30 tons per square inch applied; the rod was usually about ¼ inch diameter, and currents up to 300 amperes passed.

Results. The silica was melted to the form of a small hen's egg around the rod. When the current was increased to about 250 amperes the rod became altered to graphite, the greater the heat apparently the softer the graphite. There was no action between the silica and the carbon, the surface of the carbon remained black, and there were no hard particles in or on the carbon rod.

Other substances, such as an hydrated alumina and mixtures of alumina and silica, gave the same results.

The density of the carbon was considerably increased, in some cases from normal at 1·6 to 2·2 and 2·4; in these cases the carbon appeared very dense, much harder than the original carbon, and about as hard as the densest gas-retort carbon. No crystalline structure was visible.

The specimens were treated with solvents, and there appeared no indication of the surrounding substance having penetrated the rod; the carbon was undoubtedly consolidated by 30 per cent.

In some cases, when the material surrounding the rod was alumina saturated with oil, soft crystals of graphite exuded from specimens that had been kept for some weeks.

No. 2. Pure hydrated alumina, carbonate and oxide of magnesia, and lime all rapidly destroyed the carbon rod by combining with it, the hydrated alumina forming large volumes of gas of which it appeared to be a constituent. On account of the great diminution of bulk, no analysis was made; the gas issued from the mould explosively at from 10 to 12 tons per square inch. The alumina was found in a crystalline crust, like sugar, around where the rod had been. Hardness that of corundum, almost translucent.

No. 3. The following is the most interesting experiment of the series: On the bottom of the mould was a layer of slaked lime about $\frac{1}{4}$ inch thick, over this silver sand 2 inches, then another layer of lime of the same thickness as the former, finally a layer of coke dust, and then the plunger. With a pressure of from 5 to 30 tons per square inch in the mould, and the carbon of from $\frac{1}{4}$ to $\frac{5}{16}$ inch diameter, currents of from 200 to 300 amperes were passed.

In from 10 to 30 minutes the current was generally interrupted by the breaking or fusing of the rod, or by the action of the lime in dissolving it at the top or bottom. On opening the mould when it had cooled a little, the silica usually appeared to have melted to an egg-shaped mass, and mixed somewhat at the ends with the lime; the surface of the carbon appeared acted on, and sometimes pitted and crystalline in places; silica adhered to the surface, and beneath, when viewed under the microscope, appeared a globular cauliflower-like formation of a yellowish colour, resembling some specimens of "bort".*

After several days' immersion in concentrated hydrofluoric acid, this formation remained partly adherent to the carbon; on the surface of the carbon was a layer or skin about $\frac{1}{64}$ of an inch thick of great hardness, on the outside grey, the fracture greyer than the carbon, but having a shining coke-like appearance under the microscope.

The powder scraped off the surface of the rod has great hardness, and will cut rock crystal when applied with a piece of metal faster than emery powder. It has, under the microscope, the appearance of bort, the minute particles seem to cling together; they are not transparent as a rule, and though some such particles are found among them, it is not clear that such are hard.

When a piece of the skin has been rubbed against a diamond or other hard body, the projecting or hard portions have a glossy coke-like appearance.

A piece of the skin will continue to scratch rock crystal for some time without losing its edge. It will scratch ruby, and when rubbed for some

* The bort-like powder is not acted on by hydrofluoric and nitric acids mixed.

time against it will wear grooves or facets upon it. When a cut diamond is rubbed on the surface of the skin, it will cut through into the carbon beneath, making a black line or opening about $\frac{1}{4}$ inch long; the facet on the diamond, originally $\frac{1}{32}$ inch diameter, will have its corners evenly rounded, and its polished surface reduced to about one half its original area; the appearance of the edges is as if they had been rubbed down by a nearly equally hard substance.

The subject of the last experiment is scarcely sufficiently investigated to warrant any definite conclusions.

The substance in the several ways it has so far been tested seems to possess a hardness of nearly if not quite the first quality. The minuteness of the particles, which appear more or less cemented together, and are less cohesive after the action of acid, make it very difficult to determine their distinctive features.

The mode of formation is not inconsistent with the conditions of pressure, temperature, and the presence of moisture, lime, silica, and other substances as they appear to have existed in the craters or spouts of the Cape Diamond Mines at some epoch.

From the few experiments that have been made it appears that at pressures below 3 tons per square inch, the deposit does not possess the same hardness, though somewhat similar in appearance.

What part the lime and silica play, whether the former only supplies moisture and oxygen which combine with the carbon, or whether the presence of lime is necessary to the action, is not clear.

We may, however, observe that so far it seems as if the lime and moisture combining with the carbon form a gas or liquid at great pressure, which combining with the silica forms some compound of lime, silica, and carbon, or perhaps pure carbon only, of great hardness.

DIAMOND MANUFACTURE

With a view of ascertaining the behaviour of carbon at high temperatures and pressures, and in contact with a variety of substances, the above experiments were described in a paper to the Royal Society, June 13th, 1888.

These experiments are of interest from the fact that it was found that under certain conditions of temperature, pressure, and substance in contact with carbon, hard particles resembling a diamond were produced, which satisfied all the tests for diamond, so far as they could be applied to particles under $\frac{1}{500}$ inch in length.

At the time of reading the paper a few tests only had been applied to ascertain whether the particles found were veritable diamonds. Shortly after, however, they were examined by Professor Crookes with electrical discharge in high vacua, and appeared to him to behave in a similar manner to diamond powder.

Tests of specific gravity by immersion of the particles in borotungstate of cadmium and iodide of methylene gave a density of 3·3 to 3·5.

The particles appeared to consist of two kinds—one, irregular opaque black particles; the other, translucent plates resembling flakes of mica, generally of square or irregular shape.

When placed in a cell in iodide of methylene and projected by an electric lantern on the screen, they were clearly seen—the plates appeared to be about $\frac{1}{500}$ inch in length and of extreme thinness.

On subjecting the powder to the blowpipe all hard particles disappeared, leaving a yellowish-grey residue, but it should be stated that the powder was not previously levigated to remove the lighter portions, which would account for the residue.

SOME NOTES ON CARBON AT HIGH TEMPERATURES AND PRESSURES

Read before the Royal Society, June 27th, 1907

Following the subject of my paper of 1888 to this Society, which will be referred to in a subsequent communication, attempts have recently been made to melt carbon by electrical resistance heating under pressure, and the following is a short summary of the results of about one hundred experiments.

The procedure has been on two lines. In the first, carbon is treated in bulk in a thick tube of 8 inches internal diameter of gun steel closed below by a massive pole of steel insulated from but gas tight with the mould and above by a closely fitting steel ram packed by copper rings imbedded in grooves in the ram or by leather and steel cups according to whether solids, liquids or gases are to be contained. The bore of the mould is generally lined with asbestos and after being charged the whole is placed under a 2000-ton press, the head and baseplate being insulated and connected to the terminals of a 300-kilowatt storage battery with coupling arrangements for 4, 8, 16 or 48 volts.

It was hoped that the greater thermal and electrical conductivity of steel as compared with carbon or graphite at moderate temperatures would with the help of water jackets keep the outer layers comparatively cool, and that the increased conductivity of the central portions consequent on their higher temperature and conversion to graphite would so centralise the current on the core lying between the poles as to melt it.

Further concentration of current was obtained in the initial stages of heating by packing the central portion with carbon rods on end or by a compressed graphite core, and filling in around with coarsely broken arc-light carbon, or with wood charcoal (which is a bad conductor until highly heated).

With pressures of about 30 tons per square inch, and currents commencing at 6000 amperes, increasing up to 50,000 amperes, with about 2 volts between the terminals of the mould, the carbon rods were partially converted to graphite and firmly welded together; in the case of the graphite core the flakes were much increased in size.

The heating was in all cases limited by the melting of the steel poles and resulted in short circuits in the mould from the permeation of the asbestos by the molten iron. Neither the internal water jacketing of the poles nor the substitution of copper poles for steel have remedied this trouble.

It appears that the thermal conductivity of the carbon or graphite at or near the temperature of vaporisation is very greatly in excess of that anticipated, or that the rapid transfer of heat is caused by carbon vapour, which appears to have a great power of penetration through carbon at high temperatures. The melting of the poles and the destruction caused by short circuits which reached 80,000 amperes in the mould were not only costly to remedy, but caused contamination of the carbon from the metal of the poles and the insulating material.

In several experiments a nucleus of very soft graphite about 2½ inches in diameter was found in the centre. And in several experiments small masses of iron, highly charged with graphite, were found in varying positions among the carbon or graphite.

This method, however, would probably be more successful if carried out on a much larger scale, as for a given central temperature the transfer of heat to the poles and mould would be less, and water jackets would then prove more effective. It is, however, difficult to construct water jackets to withstand more than 30 tons per square inch, and unless made of hard steel they crush in. The maximum power of the press is 2500 tons, and with the apparatus at hand if the size of the mould was much increased the pressure in the mould would have to be decreased.

Another plan was then adopted of interposing an insulating barrier of some refractory material with a hole in it between the poles, the charge in the first instance being graphite. It was hoped that by means of electrical currents of higher potential and large volume the energy would be so concentrated on the small volume in the neck as to melt it before it had time to form carbides with the material of the barrier.

This was to some extent achieved in that the graphite in the centre was converted to a softer and more flaky nature.

In one of these experiments the barrier was formed out of a block of fused magnesium oxide, specific gravity 3·65, and the pressure in the mould, which was 4 inches in internal diameter, was in this case raised to 100 tons per square inch. The strongest steel poles were required for this pressure, also the mould of gun steel became permanently strained and required reboring after each experiment.

A current at about 12 volts at the terminals in the mould, developing about 100 kilowatts, was turned on for 7 seconds.

The initial diameter of the hole in the barrier was ⅝ inch and the thickness about ¾ inch.* This barrier was converted to magnesium carbide of a green colour to a radial depth of about ⅜ inch. Thus this magnesium oxide when heated under pressure with graphite readily forms a carbide. The

---

* The heat units delivered on to the neck being about four times that required to raise the graphite column through 5000° C., taking the specific heat at 0·5.

graphite in the centre was altered to large and very soft flakes. Neither the graphite nor the magnesium carbide contained any hard crystalline carbon.

Similar experiments were tried with carbon rods surrounded by silica, and as a guide to the temperature reached, current was turned on of just sufficient voltage to convert the rod to graphite; the mould was then set up afresh and double the voltage applied, when the rod was vaporised and disseminated throughout the molten silica, principally in the form of graphite of very small grain, very little silicon and still less silicide of carbon being formed.

Another series of experiments has been made to investigate the behaviour of vaporised carbon under fluid or gaseous pressures of about 30 tons per square inch. The general arrangement of the mould consisted of a central carbon rod with a lining of marble; in some cases the space between the rod and marble was packed with coarsely powdered charcoal.

Several compounds of carbon were treated, perhaps the most interesting being carbon dioxide. The liquid was run into the mould and a pressure of 30 tons per square inch applied. It was found that its volume diminished to about 80 per cent., due to its compressibility. Current was then passed through the rod, and the liquid must then have existed as gaseous carbon monoxide in the hotter zones.

When cooled, the liquid and gas were allowed to escape; a sample of this gas on analysis was found to contain 95 per cent. of carbon monoxide and 3 per cent. carbon dioxide, the residue consisting apparently of nitrogen.

As the pressure of 30 tons was maintained throughout the experiment, it would seem that the compressibility of carbon monoxide diminishes rapidly at such high pressures, but this experiment will be repeated and will form the subject of a subsequent paper on the compressibility of liquids and gases. Part of the central carbon was converted to graphite, and in one place there was found a nest of woolly deposited carbon, showing that under a pressure of 30 tons per square inch carbon vaporised in carbon monoxide is deposited in the form of amorphous carbon.

CONCLUSIONS

From these experiments several hundred samples have been carefully analysed. In none of the experiments designed to melt or vaporise carbon under pressure has the residue contained more than a suspicion of black or transparent diamond.

In no experiment we have made has there been any sign of the carbon becoming a non-conductor, and the impression derived is undoubtedly that soft crystals of graphite are the resulting stable form of carbon after heating to very high temperatures.

At very high temperatures and pressures graphite has a great tendency to permeate or diffuse into its cooler surroundings. It should, however, be noted that in all the experiments so far made it has been found impossible to exclude from the graphite other substances in the liquid or gaseous state.

Though in many of the foregoing experiments the molten steel of the poles became highly charged with graphite, further experiments have been made to ascertain the influence of pressure upon iron highly charged with carbon. Cores formed of iron rods, iron tubes filled with carbon or with various proportions of iron filings and lamp black, surrounded with various substances such as charcoal, magnesia, olivine, etc., were melted or vaporised and disseminated throughout the charge.

Thus iron highly charged with carbon under a pressure of 30 to 50 tons was cooled at various rates according to its proximity to the sides of the mould, the analysis showing in most cases no residue at all, but occasionally a suspicion of very minute diamond. As a further experiment, a small carbon crucible containing iron highly charged with carbon from the electric furnace was quickly transferred to a steel die and subjected, while still far above the melting point, to a pressure of 75 tons per square inch.

The analysis showed scarcely any crystalline residue and probably less than if the crucible had been cooled in water at atmospheric pressure, and as it would seem that 75 tons or even 30 tons per square inch must be a greater pressure than can be produced in the interior of a spheroidal mass of cast iron when suddenly cooled, the inference from these experiments seems to be that mechanical pressure is not the cause of the production of diamond in rapidly cooled iron.

We hope to be able to communicate further experiments on this subject during the course of next session.

I would wish to add that most of the analyses have been made by Mr J. Trevor Cart.

EXPERIMENTS ON THE ARTIFICIAL PRODUCTION
OF DIAMOND

Lecture delivered before the Royal Society on April 25th, 1918

INTRODUCTION

In this paper is given an account of experiments on the artificial production of diamond which I commenced in 1887, and carried on intermittently till the commencement of the War, when they were interrupted. Although

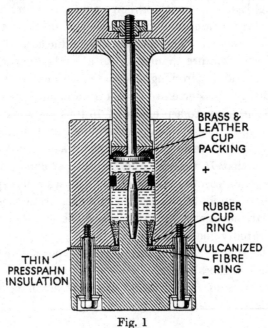

Fig. 1

the account is not as full as I could have wished, yet it is hoped that from the description of such experiments as relate to the salient features, followed by a summary of their bearings upon the research, and the conclusions at which we have arrived, a fair idea may be gathered of this research.

One reason for writing this paper at the present time has been a publication on the same subject by Otto Ruff in *Zeitschrift für Anorganische Chemie*, Vol. XCIX, pp. 73–104, May 25th, 1917, who also referred to the work of Lummer on the apparently molten aspect of the surface of the carbon of the electric arc.

In my paper to the Royal Society in 1888 were described experiments where a carbon rod heated by a current of electricity (Fig. 1) was immersed in liquids at pressures up to 2200 atmospheres, and where the liquids—benzene, paraffin, treacle, chloride and bisulphide of carbon—were found to yield deposits of amorphous carbon.

In my paper of 1907 allusion was made to experiments in liquids at a pressure of 4400 atmospheres, and to the distillation of carbon in carbon monoxide and dioxide at this pressure with similar results, also to an attempt to melt carbon at pressures up to 15,000 atmospheres, which produced soft graphite, and an experiment where a carbon crucible, containing iron previously heated and carburised in the electric furnace, was quickly transferred to a steel die, and while molten and during cooling subjected to a pressure of 11,200 atmospheres, the analyses showing less crystalline residue than if the crucible had been cooled in water.

It was also emphasised that the pressure of 11,200 atmospheres must be greater than could be produced in the interior of a spheroidal mass of cast iron when suddenly cooled, and that the inference from these experiments was that mechanical pressure is not the cause of the production of diamond in rapidly cooled iron, as had been supposed by Moissan. This conclusion appears to us in the light of our more recent experiments to be one of great importance, and it will be further discussed in this paper.

It may be well to state that, in order to facilitate a clearer view of the bearing of each experiment on the subject, they are not placed always in chronological order. The difficulty of ensuring satisfactory experiments and the elusive character of the analyses must be the excuse for the random character of some of the former. The great majority of the experiments were failures as regards results, but a few have given information that was scarcely anticipated when they were devised.

Several thousand experiments have been made and a much greater number of analyses, generally following the methods of Moissan and Crookes; the more important experiments are described at some length, and in most cases are typical of groups or repetitions of the same experiment with small variations.

The selection has been chiefly determined by their bearing on the general trend of the results of our own work and the work of others.

Those who are familiar with analyses for the detection and isolation of minute particles of diamond will know of the tendency of such particles to float, and to become lost in the frequent washings. To diminish the risk of arriving at erroneous conclusions the analyses of the more important experiments have generally been repeated several times.

EXPERIMENTS UNDER HIGH PRESSURE

In the experiments designed to test chemical reactions under high pressure, where the charge was heated by passing an electric current

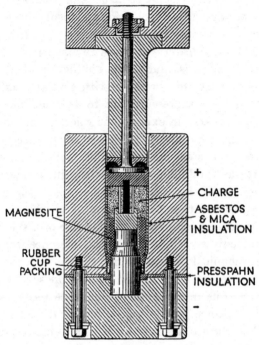

CHARGE

ASBESTOS
& MICA
INSULATION

MAGNESITE

RUBBER
CUP
PACKING

PRESSPAHN
INSULATION

Fig. 2

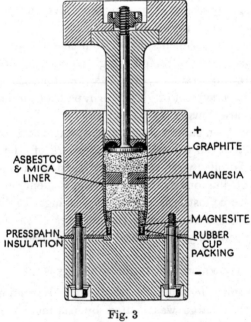

GRAPHITE

MAGNESIA

ASBESTOS
& MICA
LINER

MAGNESITE

PRESSPAHN
INSULATION

RUBBER
CUP
PACKING

Fig. 3

through a central core (Fig. 2) small residues of diamond occasionally occurred. A review of these experiments, however, indicates in most cases an association with iron, whether introduced intentionally, or present from the melting of the poles, or from other causes.

EXPERIMENTS DESIGNED TO MELT CARBON UNDER PRESSURE BY RESISTANCE HEATING

In the attempts to melt carbon under pressure by this method (Fig. 3) heat was applied for a duration of 5 seconds, sufficient in amount to melt the graphite core six times over, with the result of only altering the structure. Richard Threlfall independently came to the conclusion from his experiments at about the same time, 1907, that under 100 tons per square inch, graphite, electrically heated, remained graphite.

It appeared, however, desirable further to investigate the possibility of carbon losing its electrical conductivity when approaching its melting point, as alleged by Ludwig and others, and of thus shunting the current from itself on to the contiguous molten layers of the insulating barrier surrounding it. There had, however, been no indication of this having occurred, even momentarily; the evidence was rather that the graphite core had been vaporised and condensed in the surrounding parts of the charge, yet it was thought well to repeat the experiment with rods of iron and tungsten embedded in the core, so that should the temperature of volatilisation of the metals under a pressure of 12,000 atmospheres exceed that necessary to liquefy carbon under the same pressure, the presence of these metals might produce a different result. No change, however, occurred, though in one experiment the pressure was raised to 15,000 atmospheres.

EXPERIMENTS DESIGNED TO MELT CARBON UNDER PRESSURE BY THE RAPID COMPRESSION OF FLAME

A different mode of attack was then arranged, which would ensure that carbon should be subjected to an extremely high temperature concurrently with high pressure, obtained by the rapid compression of the hottest possible flame, that of acetylene and oxygen, with a slight excess of the former to provide the carbon.

The arrangement was as follows (Figs. 4 and 5):

A very light piston made of tool steel was carefully fitted to the barrel of a duck gun of 0·9 inch bore; the piston was flat in front, lightened out behind, and fitted with a cupped copper gas check ring, the cup facing forward; the total travel of the piston was 36 inches. To the muzzle of the gun was fitted a prolongation of the barrel, formed out of a massive steel block, the joint being gastight. The end of the bore in the block was closed by a screwed-in plug made of tempered tool steel, also with a gastight collar. A small copper pin projected from the centre of the plug to give a record of the limit of travel of the piston.

The gun was loaded with 2 drachms of black sporting powder, which amount had been calculated from some preliminary trials. The barrel in front of the piston was filled with acetylene and oxygen, with a small excess of acetylene. It was estimated that this mixture would explode when the piston had travelled about half-way along the bore; when fired the piston

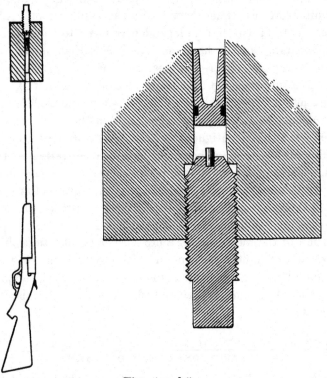

Figs. 4 and 5

travelled to within $\frac{1}{8}$ inch of the end, as had been estimated, giving a total compression ratio of 288 to 1.

Result. The surfaces of the end plug, the fore end of the piston, and the circumference of the bore up to $\frac{3}{8}$ inch from the end of the plug had been fused to a depth of about 0·01 inch and were glass hard, the surface of the copper pin had been vaporised and copper sprayed over the surface of the end plug and piston.

The end plug showed signs of compression, and the bore of the block for $\frac{3}{8}$ inch from the plug was enlarged by 0·023 inch in diameter, both deformations indicating that a pressure of above 15,000 atmospheres had been reached. A little brown carbon was found in the chamber, which was easily destroyed by boiling sulphuric acid and nitre with no residue. There was a small crystalline residue from the melted layer of the end plug, from which was isolated one non-polarising crystal, probably diamond, but too small to identify with absolute certainty.

Considering the light weight of the piston and the short duration of the exposure to heat, also the small diameter and volume of the end clearance space, the observed effects would seem to indicate that a very abnormal temperature had been reached, many times greater than exists in the chambers of large guns. There was, however, no evidence of any melting and recrystallisation of the free carbon present.

EXPERIMENTS WITH HIGH VELOCITY BULLETS

As it seemed desirable to try the effect of still higher pressures, a rifle, 0·303 inch bore, was fitted with a specially strong breech mechanism by Rigby, capable of withstanding a charge of cordite 90 per cent. in excess of the service charge.

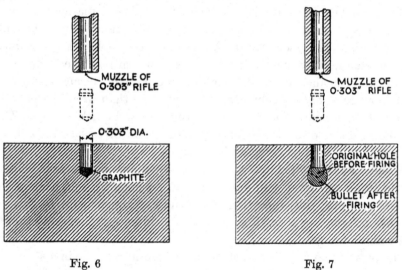

Fig. 6   Fig. 7

The gun (Fig. 6) was fixed in a vertical position on the wall of the armoured press house, with its muzzle 6 inches from a block of steel, in which a hole 0·303 inch diameter had been drilled to a depth somewhat greater than the length of the bullet, and in alignment with the bore of the gun; the trigger was pulled by a string from without. Cylindrical bullets of steel with a copper driving band were used, shorter than the service bullet, and about one-half of the weight, some with cupped noses to entrain material, some with coned noses to match the bottom of the hole in the block. The velocity with 90 per cent. excess charge was estimated to be about 5000 ft. per sec.

The substance to be compressed was placed either at the bottom of the hole when the coned-nose bullet was used, or over the mouth of the hole when the cupped-nose bullets were used. Some of the bullets were of mild steel, but those with cupped noses were of tool steel.

The substances placed in the hole included graphite, sugar carbon, bisulphide of carbon, oils, etc., graphite and sodium nitrate, graphite and fulminate of mercury, finely divided iron and fine carborundum, olivine and graphite. After each shot (Fig. 7) the bullet and surrounding steel were drilled out, and the chips and entrained matter analysed.

Several experiments were also made with a bridge of arc-light carbon just over the hole, raised to the limit of incandescence by an electric current, and the shot fired through into the hole at the moment the carbon commenced to vaporise, as observed in a mirror from without. Also an arc between two carbons was arranged just over the hole (Fig. 8) and the shot fired through it, as also through a crucible of carbon with a very thin bottom containing a little molten highly carburised iron.

Of all these experiments the only ones that yielded a reasonable amount of residue were one made with graphite wrapped in tissue paper, the bullet, however, in this case grazed the side of the hole, thus producing some molten iron by the friction; and the shots through the incandescent bridge, where again some molten metal would probably occur. The residues were in all cases exceedingly small and not more than would be produced from a small amount of iron melted, carburised and quickly cooled. There was no evidence of any incipient transformation of carbon in bulk into diamond that could be detected by analysis.

A bullet was also fired into a long hole, 0·303 inch in diameter, bored in a steel block and filled with acetylene gas, retained by gold-beaters' skin over the mouth, thus repeating the flame experiment (but in this case without oxygen) on a small scale with the intensest pressures available. The residue was nil.

The pressure on impact of a steel bullet fired into a hole in a steel block which it fits is limited by the coefficient of compressibility of the steel, and with a velocity of 5000 ft. per sec. is about 2000 tons per sq. inch. Measurements made from a section through the block and bullet (Fig. 7) showed that the mean retarding force on the frontal face, after impact till the bullet had come to rest, was about 600 tons per sq. inch.

Several experiments were made by substituting a tungsten-steel block, and a hole tapering gently from 0·303 inch at the mouth to 0·125 inch at the bottom, and using a mild steel bullet, which on entry would be deformed and a greatly increased velocity imparted to the nose. Progressively increased charges were used, and even with relatively small charges the block cracked on the second round. With the 90 per cent. excess charge, the block always split on the first shot, but this probably occurred after impact, and not till the full instantaneous pressure had been exerted, which was estimated to be greater than with the plain hole, probably over 5000 tons.

Only graphite was placed at the bottom of the hole in these latter experiments, and the analysis yielded nothing.

EXPERIMENTS ON PRESSURE IN CAST IRON WHEN COOLED

It has been generally assumed that iron rich in carbon expands on setting, and that this supposed property is a contributory cause in the formation of diamond.

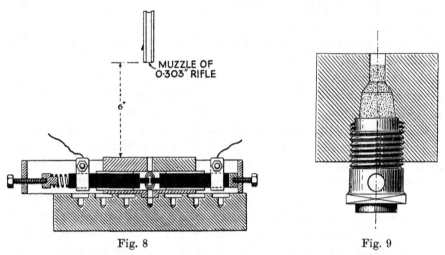

Fig. 8   Fig. 9

Several experiments were made by pouring iron saturated with carbon from the electric furnace through a narrow git into a very massive steel mould, closed at the bottom with a breech screw (Fig. 9). When cold, the breech screw was easily removed, and there was no sign of any appreciable pressure having come on the threads. Not being sure that, because of capillarity, the corners of the mould had been quite filled, a steel mandril was, immediately after pouring, forced down the git-hole by a press giving a fluid pressure in the mould of 75 atmospheres. The observed pressure on the breech screw appeared not to have exceeded this pressure. Highly carburised iron, therefore, does not expand with any considerable force on setting.

The reason why a lump of cast iron thrown into a ladle of molten metal first sinks to the bottom and soon rises and floats on the surface is probably that cast iron is about seven times stronger in compression than in tension.

Therefore when a sufficiently thick layer of the cold metal has been heated the interior is torn asunder by the expansion of the outer skin, and the specific gravity of the whole mass is diminished. (See Mr Wrightson's paper "On Iron and Steel at High Temperatures", with discussion, *Journal of the Iron and Steel Institute*, No. 1 for 1880.)

We may therefore safely conclude that when iron is suddenly cooled, the only compressive bulk pressure that is brought to bear on the interior is that arising from the contraction of the outer layers after setting, and with highly carburised iron this can only be small because of the low tensile strength of the metal.

GASES EJECTED FROM CAST IRON ON SETTING

As bearing upon the question of the possibility of the occluded gases playing a part, Moissan was the first to observe that spherules or small spheres of iron with cracks and geodes never contained diamond. We have made experiments by pouring highly carburised iron, alloys and mixtures on to iron plates, the cooling taking place from one side only, and under such conditions no diamond results; in fact it only occurs when the ingot or spherule is cooled on all sides nearly simultaneously, so that an envelope of cold metal is formed all over before the centre sets.

Since my paper in 1907, the experiment of heating iron in a carbon crucible and transferring it to a steel die and subjecting it to 11,200 atmospheres pressure has been repeated, and it has been found that if the iron is allowed to set before the pressure is applied the amount of diamond is much greater than if pressed when very hot and molten, and that it is then about the same as when the crucible is cooled in water. The only reason that suggests itself to account for this is, that when pressure is applied while the iron is very hot some of the latter permeates the carbon of the crucible, and because of the greater specific heat and lesser conductibility of the carbon, the iron next to and in the carbon remains molten after the ingot has been cooled by direct contact with the steel cup on the face of the plunger. Thus, when cooling, the occluded gases have a free exit from the ingot, through the molten metal (which is pervious to gas) into the carbon of the crucible, and are not retained in the ingot to the same extent as when it is set and enclosed in an envelope of colder iron impermeable to the gases before pressing.

The experiments of Baraduc Muller (*Iron and Steel Institute, Carnegie Scholarship Memoirs*, 1914, p. 216), on the extraction of gases from molten steel, showed that steel is permeable to gases down to 600° C.

OTHER EXPERIMENTS

The action of water on carbide of calcium, and of concentrated sulphuric acid on sugar for 6 hours under pressure of 30,000 atmospheres, was tried; in both cases amorphous carbon was formed and no diamond.

Hannay's experiments were repeated, where paraffin and Dippel's oil with the alkali metals, especially potassium, were sealed in steel tubes and subjected to a red heat for several hours. The analysis gave no diamonds; in fact it became apparent that when hydrocarbons or water were relied on to produce pressure, the latter could only exist for a short time at the commencement, for when a red heat was reached the hydrogen escaped through the metal, and the oxygen combined with the steel.

We did not analyse the steel tubes themselves. Many experiments were, however, tried with central heating under the press at 6000 atmospheres, and nothing was obtained of interest with the substances used by Hannay, unless, as previously mentioned, some iron was present. Friedlander's experiment was repeated, where a molten globule of olivine, in a reducing flame, or with carbon added, was stated by him to contain minute diamonds. An experiment was made with molten olivine in a carbon crucible in a wind furnace stirred with a carbon rod, with and without an electric current passing between the rod and crucible.

Many experiments were also tried at 6000 atmospheres under the press with central heating with olivine associated with carbon, hydrocarbons, bisulphide of carbon, water, etc., also with blue ground from Kimberley instead of olivine. The results of the analyses were in all cases negative, except occasionally when metallic iron was present. Thus in some cases the olivine or blue ground was partially smelted by the heating carbon rod or by the associated hydrocarbons, etc., when such were added, and iron globules were formed. In these, diamond was occasionally found when cooling was rapid and they were centrally situated in the charge.

Very quick cooling. To test the action of very quick cooling a carbon crucible of 2 inch internal diameter charged with iron, sugar carbon, 2 per cent. silicide of carbon, well boiled by resistance heating under atmospheric pressure and 2 per cent. of iron sulphide added, was quickly placed on asbestos mill-board resting on a steel table frictionally held in the bore of the 4-inch mould, below being placed 2 lb. of carbon dioxide snow, and the plunger quickly brought down by the press, subjecting the whole to 6000 atmospheres pressure. When taken out the crucible was intact, the contents had divided into a lower portion consisting of a large grained crumbling mass of graphite admixed with granules of very hard iron, in the centre a rounded pillar of white iron equally hard. The cooling seemed to have been unusually rapid.

The experiment was repeated, the crucible being charged with iron, sugar carbon, 5 per cent. manganese, 5 per cent. cobalt, 2 per cent. silicide of carbon, boiled, and 2 per cent. iron sulphide added.

It was also repeated with water instead of carbon dioxide snow. The result of all these experiments was similar to the first. No diamond was found in any part.

An experiment which seemed to give practically instantaneous cooling was as follows: A small carbon crucible containing iron, with traces of silicon, aluminium, calcium, magnesia and sulphur, was floated on a carbon block on a bath of mercury, all contained in a vessel exhausted to 2 mm. absolute. The crucible was heated by an arc from an upper carbon, the holder passing through a stuffing-box. When the crucible was sufficiently hot and the contents carburised, the upper carbon was thrust down, submerging the crucible under the mercury; the cooling was almost explosive and instantaneous—the finely divided iron and graphite on analysis yielded no diamond.

Extremely rapid cooling does not, therefore, seem to be a direct cause in the production of diamond.

EXPERIMENTS AT ATMOSPHERIC PRESSURE

A convenient method of studying the effect of the association of other elements with iron on a small scale uncontaminated by the vapours of a furnace lining suggested itself, and a series of .experiments was made as follows: A deep iron dish was packed tightly with Acheson graphite with a slight dimple in the centre to hold the ingot; above, graphite was filled in loosely to a depth of half an inch covering the ingot. An arc was struck by a carbon on to the ingot submerged in the loose graphite. When the iron was well boiled the surrounding graphite with the ingot in it was dug out entire and thrown into a bowl of mercury covered with water.

The results showed that, using ordinary mild steel, no diamond ever occurred on analysis, but that a small percentage of silicon is absolutely essential; small percentages of aluminium, magnesium, calcium, one or all are important; sulphur, manganese, and cobalt increase the yield, nickel appeared to be a disadvantage. An alloy of iron and 10 per cent. manganese, 10 per cent. cobalt, and 5 per cent. silicon gave out much gas when cooled slowly, and on quick cooling in water and mercury most of the spherules were burst and shredded.

Finally about 1 to 3 per cent. of the other elements added to iron appeared to give the best results and the spherules were not then burst.

An experiment was made by letting the ingot remain in the bed till it had quite set, hard enough to handle with the iron spoon, and then, cooled in water and mercury. It gave a fair diamond residue.

EXPERIMENTS ON THE CONVERSION OF DIAMOND TO GRAPHITE

A clear octahedral diamond was placed in a small carbon crucible and packed loosely with Acheson graphite and heated for 10 minutes to about 1400° C. The diamond was coated with a firm layer of graphite.

After two prolonged treatments with fuming nitric acid and potassium chlorate, alternating with boiling sulphuric acid and nitre, the opaque coating was removed and there remained a blackish translucent skin. When fractured the interior was unaltered and perfectly transparent.

A piece of bort somewhat laminated, after the same treatment, showed the laminations separated by cracks starting from the outside. Upon breaking, the interior surface of the fissures showed an incipient change to graphite, but less rapid than on the outside surface. There was a sinuous pitting, deepest near the outside and diminishing inwards. The substance of the bort between the fissures was unaltered.

The change of diamond to graphite under the conditions described is gradual, the surrounding gases, carbon monoxide, carbon dioxide, nitrogen, hydrogen, and also vapour of iron (as an impurity in the graphite) singly, or collectively, probably play a part, and further investigation as to this seems to be desirable.

Sir James Dewar, in 1880, heated a diamond in a carbon tube to a temperature of 2000° C., while a flow of pure hydrogen was maintained through the tube. The diamond soon became covered with a coating of graphite (*Proceedings of the Royal Institution*).

A clear diamond plunged into molten iron saturated with carbon at about 1400° C. for 5 minutes was deeply pitted. When removed from the iron small globules of iron adhered to the surface and the pits appeared to occur at these spots.

A clear diamond was disintegrated by cathode rays, the temperature by pyrometer being 1890° C., the splinters were quite black and opaque, but after several prolonged treatments with fuming nitric acid and potassium chlorate, alternating with boiling sulphuric acid and nitre, the coating that remained was a dusky grey, but semi-transparent, the gas present being chiefly hydrogen. (Paper by Parsons and Swinton, January 16th, 1908, *Roy. Soc. Proc.* A, Vol. LXXX.)

In this latter experiment the surface action appeared to be much less in proportion to the incipient change of the under layer to graphite, and the impression is that at 1890° C. the temperature of bulk transformation is being approached, also that carbon monoxide, carbon dioxide, nitrogen, hydrogen, and iron, one or more, act as catalysts in the change of diamond to graphite.

EXPERIMENTS ON THE OXIDATION OF ALLOYS OF IRON WHEN MOLTEN

Iron was melted in a carbon crucible and highly carburised; when it had somewhat cooled, the other elements were added, in small percentages of aluminium, silicon, calcium, magnesium, manganese, iron sulphide, collectively and in some cases singly; the crucible was then removed from the furnace and superheated steam blown through a carbon tube into the metal; energetic action took place and much heat was evolved; on analysis, after destroying the graphite, a bulky transparent crystalline residue remained.

With aluminium alone the crystals were chiefly crystallised alumina, and with the other elements the spinels and other crystals were produced; all were transparent and colourless, but when chromium was added some rounded crystals occurred resembling pyrope. When submitted to sulphur dioxide and carbon dioxide the result was the same, but less residue was produced. Under the microscope there appeared to be a small proportion of very small crystals like diamond; these burnt in oxygen. When the bulky residue was placed in a test tube with the double nitrate of silver and thallium, and the density adjusted so that a diamond floated midway between the top and bottom, there collected into its immediate neighbourhood after a time an amount of the small crystals which was estimated to be about 5 per cent. of the total residue.

One prolonged treatment of hydrofluoric acid had no apparent effect on the bulky residue, and it required so many treatments to destroy it that we failed to isolate the very small particles whose size did not exceed $\frac{1}{20}$ mm.; they were probably lost by flotation. These experiments were repeated many times with the same result, but they merit further investigation, with steam under high pressure and conditions favourable to the formation of larger crystals.

Note. Marsden observed in silver the association of black diamond with crystalline alumina, silicide of carbon, etc., *Roy. Soc. Proc.* 1880.

EXPERIMENTS IN VACUO

The presence of diamond in some meteorites suggested a series of experiments under various degrees of vacuum up to the highest obtainable.*

It is probable that some meteoric matter may have been melted by collision or ejected into space in a molten state and cooled by radiation, and that under such conditions the absence, or diminution, of occluded gases might be a factor conducive to the crystallisation of carbon.

One of the 4-inch diameter pressure moulds (Fig. 10) was used in a

---

* Also an impression suggested itself in 1907 that hydrogen had an adverse effect on the formation of diamond.

preliminary experiment as the container. The crucible was turned out of a 1½-inch carbon rod, and so formed on a stem that the electric current heated the bottom and sides equally. The cover was similarly formed and its holder was electrically connected with the container, but free to move vertically and to rest its weight on the crucible, electrical connection to the container being made by a layer of brass or iron turnings resting on the holder. A current of 1000 amperes at 16 volts sufficed, and the temperature was observed through a glass window at the side of the container.

The crucible was charged with reduced iron and lampblack. The Geryk pump evacuated the container to ⅜ inch mercury absolute; current was turned on for 15 seconds, the vacuum fell to 3 inches, when it had risen

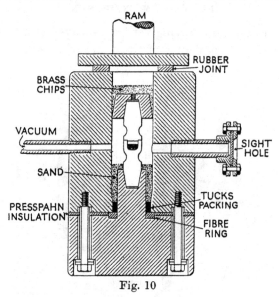

Fig. 10

again to ⅜ inch current was again turned on. This was repeated three or four times, finally current was applied for 30 seconds and the vacuum again fell to 3 inches. The gas was drawn off and collected, it amounted to a total of ½ gallon at atmospheric pressure and consisted of 95 per cent. carbon monoxide, 1 per cent. hydrogen, 2 per cent. hydrocarbon, 2 per cent. nitrogen.

The carbon which formed the crucible and cover contained a large percentage of silica, but the carbon monoxide was produced chiefly by the action of sand (of which there was a thick layer on the bottom of the container to protect the insulating joint from iron spilled from the crucible) on the carbon of the stem of the crucible. About one half of the iron had been evaporated, and there remained an ingot about the size and shape of a broad bean. It contained rather large graphite crystals and was easily broken. The analysis gave the largest residue of diamond in proportion

to the amount of iron of any of our experiments, the largest crystals being
0·7 mm. in length.

This experiment was repeated several times with the same result. The
time of cooling of the crucible, from switching off the current to the
temperature of setting, was 15 seconds, and probably sufficiently rapid to
allow a skin to be formed around the ingot before the centre was solidified, for the configuration of the crucible and cover was such as to ensure
nearly equal and simultaneous cooling on all sides of the ingot. At the
time, vacuum was erroneously thought to be the chief contributory cause
and not the presence of carbon monoxide in large proportion.

HIGH VACUUM EXPERIMENTS

The molecular pump not having yet been evolved, a powerful pumping
system was arranged, consisting of three steam-jet exhausters in series,

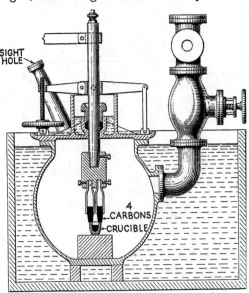

Fig. 11

the last ejector of the series discharging into a jet condenser with separate
air and water pumps, the former assisted by a steam jet. The two steam-
jet exhausters nearest to the exhausted chamber were fed with highly
superheated steam at 200 lb. pressure, and the suction pipe to the chamber
was 4 inches in diameter; the chamber, 2 feet 6 inches diameter of
spherical shape (Fig. 11). A vacuum of $\frac{1}{8}$ mm. absolute could be reached.

The crucible was placed on a large block of carbon, resting on the base
of the chamber, and forming the bottom pole. The cover was insulated
from the chamber, and through an oil-sealed gland passed a 2-inch brass

rod, carrying a crown holder, with four 2-inch carbons which rested on the lip of the crucible for resistance heating. An observation window was placed at the apex of a long iron cone, projecting from the side of the cover, which gave a good view of the crucible and its contents. The whole of the chamber was submerged in a tank of water, up to the level of the gland in the cover.

Iron and iron alloys were boiled and allowed to cool slowly by radiation, or were rapidly quenched by admitting water through a large valve from the tank into the vacuum vessel. The iron and carbon vapour from the boilings deposited dust and globules on the cover, and on the sides and bottom of the chamber. A very small diamond residue generally resulted from the small iron globules, and also from the dust, but never anything from the ingot remaining in the crucible.

In several experiments water was admitted, which played directly on the crucible, the upper carbons resting on the rim prevented its upsetting by the force of the water, and still there was no residue. In one experiment the carbons were lifted and the charge flowed out, forming spherules of varying size in the water. There was a very small diamond residue from these spherules.

In one experiment a crucible was filled with iron and carbon and closed by a tight carbon cover, a hole bored in the side of the crucible, a massive block of iron placed close opposite the hole and the crucible boiled, the vacuum being under 1 mm. No crystallised residue was found in the deposit on the iron block from this high velocity jet of vapour of iron and carbon.

In another experiment a powerful electromagnet was provided with poles to give a concentrated field, and an arc struck between two carbons was arranged to burn within this field and regulated from without by hand. There was an iron block upon which the arc directed by the field could play and condense its carbon vapour. The analysis gave no diamond.

It was thought that the vapour from boiling iron saturated with carbon might, by the action of bisulphide of carbon, cause a crystalline deposit, but all the experiments to this end yielded no results.

EXPERIMENTS UNDER X-RAY VACUUM

Experiments were made under X-ray vacuum in a new chamber of cast iron with very thick walls to absorb the heat, exhausted through an 8-inch diameter suction by a large molecular pump alongside, in series with a dry, high speed, two stage, pump, 12-inch diameter pistons, and last of the series a 3-inch + 2-inch compound Fleuss. The crucible was resistance-heated as before (Fig. 12). No diamond was produced in any of these experiments, except in those where iron, sand, and other elements, with or without sulphur, were first heated and well boiled in the carbon crucible at atmospheric pressure, and after cooling transferred to the vacuum furnace and re-heated by resistance under X-ray vacuum; violent ebullition occurred owing to the liberation of occluded gases, and many iron spherules were ejected, which cooled by radiation and conduction where they fell; diamond was found in these, which burnt in oxygen, but no diamond was ever found in the ingot remaining in the crucible.

It occurred to us to try the effect of great mechanical pressure accompanied by heat upon small particles and powders, the interstices being exhausted to a high vacuum.

Several experiments were made in the press under a mass pressure of 3000 atmospheres.

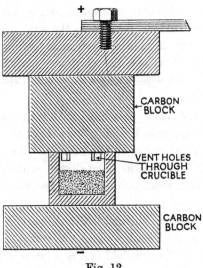

Fig. 12

A layer of cast-iron turnings resting on a layer of carborundum grit, the exhaustion being effected through a hole in the side of the mould covered by a perforated steel plate within the layer of grit, heat was applied as usual by a central carbon rod.

Analysis yielded some thin crystal plates from the grit which had lain in the line between the cast iron and the suction outlet at the grid, and also from the layer of grit which had lain against the cast-iron turnings which had become heated but not melted by the central carbon rod.

To ascertain the cause of the occurrence of these plates, experiments were made, without bulk pressure, on the concentrated action of the gases given off from cast-iron turnings heated up to a good red, and drawn by a high-vacuum pump through carborundum grit placed in a silica tube heated by a gas burner at the centre of its length to dull red. These yielded similar crystal plates.

Control experiments showed that no similar plates existed in the untreated grit.

It was also found that the cast-iron turnings would not produce this effect on a second heating unless they had been subjected to CO at atmospheric pressure for some hours. Carbon monoxide, sulphur dioxide, cyanogen, hydrogen, nitrogen, oxygen, nitric acid gas, chlorine, ammonia, ammonium oxalate vapour, ammonium chloride, acetylene, or coal gas, produced no plates.

These plates resemble diamond very closely in appearance and form of crystallisation, they do not polarise, and some have triangular markings; they will not, however, burn in oxygen at 900° C., and are completely destroyed by chlorine purified from oxygen and water vapour at 1100° C.; their specific gravity is about 3·2, they are therefore not diamond.

Note. Recent experiments have shown that carbon monoxide passed over molten iron sulphide and then over carborundum grit below red heat at atmospheric pressure also produces these plates, and that if coal gas is substituted for carbon monoxide no plates are formed. Also that only a few of the grains produce plates.

The composition of the grains is

Carborundum	...	...	...	36·56
Iron oxide and alumina	...	...	44·09
Lime ...	...	...	...	...	10·45
Magnesia	...	...	...	...	5·57

SUMMARY OF EXPERIMENTS AND CONCLUSIONS

The experiments have shown that all the hydrocarbons, chlorides of carbon, and oxides of carbon tested, deposit amorphous carbon or graphite on a carbon rod electrically heated at any pressure up to 4400 atmospheres, and in a few experiments up to 6000 atmospheres; and that at 15,000 atmospheres carbon and graphite electrically heated are either directly transformed into soft graphite or are first vaporised and then condensed as such.

While the experiment of rapidly compressing a mixture of acetylene and oxygen with the production of temperatures much in excess of that necessary to vaporise carbon, accompanied by a momentary pressure of about 15,000 atmospheres, confirms the conclusion that the negative results obtained in the attempts to convert graphite into diamond by electrical heating are not due to lack of temperature; on the other hand, the presence of minute crystals in the molten layer of the steel of the end of the barrel subjected to high gaseous pressures of carbon monoxide, carbon dioxide, and hydrogen appears to be connected with the other experiments bearing upon the inclusion of gases in metal as a factor in the production of diamond.

The experiment of firing a high velocity steel bullet with cupped nose through vaporising carbon into a hole in a block of steel has tested the effect of a momentary pressure of about 300,000 atmospheres on carbon initially near its melting point, and probably raised by adiabatic compression by another 1000° C.

The fact that only a very few minute crystals resembling diamond were produced (probably from the iron) raises the question as to whether the duration of the pressure is sufficient to start a transformation of graphite to diamond which can be detected by analysis. We have distinct evidence that, with iron as the matrix, the time is sufficient to form very small crystals which can be identified with some certainty. It therefore seems reasonable to conclude that there was no incipient transformation in bulk, and that however long the pressure of 300,000 atmospheres were applied, it is extremely doubtful if any change would occur.

The pressure of 300,000 atmospheres is between one quarter and one half that obtaining at the centre of the Earth, but vastly greater pressures exist at the centre of the larger stars, and are produced by the collision of large bodies in space; these pressures are many thousands of times greater, and whether they would effect the change it is impossible to predict. On the other hand, a heating effect on large masses of iron might be produced by collisions, and owing to the heat generated by adiabatic compression of the central portions, some of the mass would be melted and subsequently cooled on release of the pressure, so that if heating and cooling under pressure are alone necessary for the production of diamond large stones might result. These considerations, though of interest as bearing upon the presence of diamonds in meteorites and also indicating a possible origin of natural diamond, are of no practical value to us because the pressures required are entirely beyond our reach. There are, however, other considerations arising out of the experiments of Marsden, Moissan, and Crookes, as well as our own, which seem to give some hope of solutions of the problem at issue which lie within the means at our disposal.

A repetition has been made of many of the experiments in which diamond is claimed to have been produced. These have given negative results in all cases except where iron has played a part, as for instance with olivine, when being partly reduced by carbon or a reducing flame, small spherules of iron are produced and may, if the mass is quickly cooled, be found to contain diamond.

The repetition of Moissan's experiments under a variety of conditions and pressures has not only confirmed his results but has thrown, it is hoped, additional light on the causes operating to produce diamond in iron.

The experiments under high pressure in steel moulds, where heating of the charge was effected by a central core through which current was passed, enabled Hannay's experiments with Dippel's oil to be tried under much higher pressures, and more thoroughly than is possible with steel tubes in a furnace.

The Appendix* gives some indication of the many substances and chemical reactions tested. The results were chiefly negative. The few that were favourable were generally attributable, as has been said, to the presence of iron. It was noticed that the iron seldom contained diamond unless when so situated in the charge as to cause equal cooling on all sides, and it will be remembered that the experiments under atmospheric pressure showed this condition to be essential for the formation of diamond.

In some of the experiments of this group considerable gaseous pressure existed, up to 6000 atmospheres, but it is doubtful if in these the right kind of gas was present or a sufficiency of heating or carburisation of the iron occurred. On the whole, therefore, it would appear that all, or nearly all, the chemical reactions as such, under pressures up to 6000 atmospheres, have given negative results.

The experiments on very rapid cooling would seem to dispel the theory that carbon can be caught in a state of transition, and to lead us to the conclusion that quick cooling is not in itself a cause of the occurrence of diamond in rapidly cooled iron.

Moissan observed that when the spherules of granulated iron were cracked, or contained geodes, no diamond was ever found in them, and he attributed this to want of mechanical pressure. The experiments we have made not only corroborate this fact, but they tend to show, we think conclusively, that the cracks in the spherules act by allowing a free passage for the occluded gases to escape, and the geodes by providing cavities in which the gases can find lodgment without much gaseous pressure occurring in the metal.† Further, the experiments have shown that iron when it sets does not expand with appreciable force, and that the only compressive forces that are brought to bear on the interior are those arising from the contraction of the outer layers.

Our experiments further show that when a crucible of molten iron is subjected to pressure more than three times as great as can be produced by these contractile forces, the yield of diamond is not increased. On the other hand, when the conditions of the experiment operate to imprison the occluded gases, then the yield of diamond is about the same as if the

---

* Not reproduced.

† Conversely they may act to allow gases to enter the metal.

crucible had been plunged into water, while if the conditions are such as to allow a free passage through the skin of the ingot, the yield is at once diminished, even though the bulk pressure on the ingot is the same.

The experiment, on compressing acetylene and oxygen, has shown that minute crystals, probably diamond, are produced almost instantaneously in the molten surface of metal exposed on one side to gases consisting of carbon monoxide, carbon dioxide, and hydrogen at very high temperature and at 15,000 atmospheres. Sir William Crookes' experiment described in his lecture before the British Association at Kimberley in 1905 is somewhat analogous; cordite with a little additional carbon was fired in a chamber, the pressure reaching 8000 atmospheres, a few crystals of diamond were found and isolated; this result Crookes attributed to the melting of the carbon under the temperature of explosion and crystallisation under the pressure on cooling.

Under the conditions of the experiment there would be a considerable amount of the surface of the chamber melted and swept into the products of the charge by the turbulence of the explosion, and the spherules of iron would thus be carburised and cooled while still under heavy pressure.

In the acetylene-oxygen experiment there is a molten surface with reducing gases on one side at high pressure, and on the other metal impervious to gases. In Crookes' experiment the globules of metal are surrounded by gases at high pressure. In both cases the metal has solidified with the occluded gases imprisoned by the high external gaseous pressure, for we have seen that the pressure of occluded gases in highly carburised iron when quickly cooled cannot exceed about 1000 atmospheres.

The experiments under vacua from 75 mm. up to X-ray vacua have shown generally that as the vacuum is increased the yield of diamond in the crucible is diminished, and that below 2 mm. none has been detected. But when alloys previously boiled at atmospheric pressure are quickly heated up under high vacuum violent ebullition takes place, from the large volume of gases liberated, and some of the contents are ejected into the vacuum chamber before they have had time and sufficient temperature to part with their occluded gases, and diamond occurs in the spherules so ejected.

The gases occluded in cast iron which are given off when heated *in vacuo* have been investigated by H. C. Carpenter and others, and the relative amounts of the constituents are found to vary widely according to the previous heat treatment and the nature of the gases in contact with the metal while molten and during cooling; they are carbon monoxide and carbon dioxide, hydrogen and nitrogen.

H. C. Carpenter (*Journal of Iron and Steel Institute*, 1911) states that, when heating up a bar of cast iron *in vacuo* in a silica tube, "After the twenty-fifth heat it was noticed that in the water-cooled areas of the quartz tube a lustrous black ring had formed. On being strongly heated, some of this, evidently carbon, burnt off, leaving a white film, presumably silica. This seems to show that a volatile silico-organic compound, containing carbon, hydrogen, and silicon, was evolved from the iron on heating".

It would appear from our experiments that probably a ferro-silicon carbonyl is given off from the iron, for, as has been said, we observed a corrosive action on carborundum by the gas evolved from iron borings at red heat under a high vacuum, and the same action was produced by gaseous ferro-carbonyl, and also by carbon monoxide, previously passed over molten iron sulphide at atmospheric pressure.

Let us consider what happens in an ingot or spherule when rapidly cooled simultaneously on all sides. It is first surrounded by a thin coat of solidified metal which, below 600° C., is impervious to gases. As the coat thickens layer within layer, more and more gas is ejected by the solidifying metal, and its semi-solidified centre, still pervious to gas, receives the charge. As this process progresses the pressure may rise higher and higher, though there may be a limit to the pressure against which the metal is able to eject gas when setting. All we, however, know is, that the mechanical strength of the ingot or spherule places a limit of about 7000 atmospheres on the gaseous pressure, and, as we have already mentioned in the case of some iron alloys, most of the spherules are split or shredded, with an appearance consistent with this view.

Crookes' microscopical examination of diamonds with polarised light supports this view. In his lecture at Kimberley, in 1905, he states: "I have examined many hundred diamond crystals under polarized light, and with few exceptions all show the presence of internal tension.

"On rotating the polarizer, the black cross most frequently seen revolves round a particular point in the inside of the crystal; on examining this point with a high power we sometimes see a slight flaw, more rarely a minute cavity. The cavity is filled with gas at enormous pressure, and the strain is set up in the stone by the effort of the gas to escape."

It seems therefore probable, or indeed almost certain, from the accumulated evidence, that the chief function of quick cooling in the production of diamond in an ingot or spherule is to bottle up and concentrate into local spots the gases occluded in the metal which, under slow cooling, would partially escape and the remainder become evenly distributed throughout the mass.

As to the condition in which the gases exist within the iron at temperatures above 500° C. little is known, though at 200° C. and at 180 atmospheres Mond has shown that iron penta-carbonyl is formed. The intimate contact between the occluded gases and other elements, metals or carbides, must favour complex interactions as cooling takes place. Such actions might be concentrated by the heat flow across the metal on quick cooling.

It appears probable that concentration of gaseous pressure causes certain reactions which bring about an association of carbon atoms in the tetrahedral form—against their natural tendency to assume the more stable form of graphite.*

The necessity of subjecting the iron to a temperature above 2000° C. before cooling would seem to imply the necessity of carbides of the other metals, such as silicon, magnesium, etc., being present to insure the necessary chemical reactions with the gases at high pressure within the ingot.

In reviewing all our experiments, the greatest percentage of diamond occurred when the atmosphere around the crucible consisted of 95 per cent. carbon monoxide and 1 per cent. hydrogen, 2 per cent. hydrocarbons, 2 per cent. nitrogen, the mean pressure in the vessel being about 1 inch absolute of mercury. The weight of diamond we estimated to be about $1 \div 20,000$ of the weight of the iron. If we, for the moment, assume a volume of carbon monoxide at atmospheric pressure equal to 0·69 that of the iron, the weight of carbon contained in it equals that of the diamond.

For the following reasons it would appear that the formation of diamond in rapidly-cooled iron takes place when it is solid or in a plastic condition, or even at a still lower temperature. The rapid pitting of a diamond in highly carbonised iron just above its melting point is so pronounced that the largest diamond hitherto produced artificially would be destroyed in a second or two if the iron matrix were molten. The production of diamond was obtained in an ingot rapidly cooled after it had set sufficiently hard to be handled in a spoon. A similar result was obtained in the case of a crucible placed in the die and subjected to 11,200 atmospheres pressure after the contents had set. Moissan found the diamonds to occur in the centre of the ingots both in the case of iron and also of silver.

It has been seen that iron is permeable to carbon monoxide and hydrogen at temperatures above 600° C., and there appears to be no reason why the concentration of the occluded gases should not take place within the mass as effectively at 600° C. as at higher temperatures, provided that they cannot escape. The most probable temperature, however, may be the point of recalescence at 690° C.†

---

* It also appears that the conditions may operate to the exclusion of some gas or element inimical to the formation of diamond from certain parts of the metal, viz. the graphite liberated and the cooled metal of the outer layers may absorb some gas or element from the inner portion of the ingot and leave none for the central portion.

† These conditions may also operate to exclude some gases from certain portions of the metal.

It would appear that the function of the impervious metal coating thrown around the ingot by quick cooling might be better effected by gas of the same composition as that which the metal ejects on cooling, the pressure being sufficient to ensure that the gaseous pressure around the ingot shall be equal to, or greater than could occur on quick cooling. Such a substitution might result in a larger gaseous content and a larger proportion of the ingot being brought into a suitable condition for the formation of diamond, and the yield might thereby be increased. Some gradations of temperature might still be found necessary to concentrate the reactions. It seems, however, probable that the rate of cooling might be so much prolonged as to obtain much larger crystals and a larger total yield.

The presence of crystals of silica, alumina and magnesia and the spinels, and pyrope associated with diamond in rapidly cooled iron alloys, and also when oxidised by steam and some other gases, appears to have a bearing upon the presence of similar crystals usually found in association with diamond, and to be compatible with the conclusions of Bonney that eclogite is the parent rock of the diamond in South Africa. It seems probable that both the eclogite and the diamond may have been crystallised nearly simultaneously from an iron alloy.

Moissan, after a recital of the geological conditions existing in the South African pipes (see *Four Electrique*, p. 115), came to the conclusion that diamond was not a vein mineral, but must have been evolved in the midst of a plastic mass; and he concludes that iron at high pressure must have been the matrix. Our experiments, however, seem to show that bulk pressure on the metal does not play a part, but that the previous heat treatment, the impurities in the iron and the condition of the gases within the metal, are the important factors.

It is interesting to note that in the best experiments the yield of diamond in rapidly cooled iron has reached $1 \div 20,000$ of the weight of iron, whereas the weight of diamond obtained from the blue ground of the South African mines is only $1 \div 5,400,000$. This comparison appears to be confirmed by the relative rarity of microscopic diamonds we have found in the many analyses we have made of blue ground and of the conglomerate from Brazil.

Thus in cooled iron there may be more than 270 times as much diamond as exists in the bulk average of blue ground.

PART III
APPENDICES

APPENDIX A

THE PARSONS AUXETOPHONE

By A. Q. CARNEGIE*

Amongst Sir Charles Parsons' many inventions, the auxetophone—an instrument for augmenting sounds—was not only of great intrinsic interest but occupied a great deal of his thoughts and afforded him many nights of relaxation from turbine problems. As far as the Editor has been able to ascertain, Sir Charles never gave a public lecture on the instrument, but it was "demonstrated" at a Conversazione of the Royal Society on May 13th, 1904, and before the Northern Scientific Club at Newcastle-upon-Tyne about the same time. In 1906 Sir Henry Wood used it in his orchestra and a few years later Van Biene—of *Broken Melody* fame—applied it to his 'cello when playing as soloist in a concert conducted by Sir Landon Ronald at Queen's Hall.

The Editor felt, therefore, that an account of the instrument should appear with these Collected Papers, and was fortunate enough to secure the following notes by the late Mr A. Q. Carnegie who for several years interested himself in the auxetophone.

In 1901/2, when the phonograph and gramophone were just coming into popularity as home musical instruments, the gramophone was of the type illustrated in the well-known picture "His Master's Voice", and Sir Charles Parsons realised that this instrument held out great possibilities for improvement.

The art of making records had been well established and had been developed almost to the limit imposed by the method of direct mechanical recording then and for many years afterwards in use. A large selection of excellent records made by famous and other artists were obtainable, but the volume and musical quality of reproduction left much to be desired.

Parsons, deciding that the vibrating diaphragm, and the sound-box generally, were superimposing their individual characteristics and limits on reproduction, abandoned the telephone principle, and turned to the organ pipe, or mechanically operated syren principle. He thought it ought to be possible to produce good quality sound by generating controlled waves of air pressure and admitting them directly to the trumpet.

He formed a mental picture of a needle-operated air valve vibrating in sympathy with the sound waves on the record disc, thus acting as a mechanical equivalent of the vocal chords in the production of the human voice.

He found that Edison had proposed in 1877 the use of an air relay and valve moved by a diaphragm operated by sound, the air from the valve operating a second diaphragm, and this in its turn operating a microphone for the purpose of intensifying sound. Since that date other inventors had proposed the use of air relays, but no satisfactory results appeared to have been obtained.

* Deceased May 5, 1934.

To Parsons, the fact that an apparently sound principle had been proposed and dropped was nothing less than a challenge and, following his invariable practice, he made careful calculations of the dynamical forces, velocities of sound waves and their amplitudes, and the velocities of transmission of sound waves in metals in the forms which would have to be used in such instruments. He also measured the velocities of flow of air and gases through small orifices and, having studied the whole problem in his own uncanny way, decided that none of the previous experimenters had understood it sufficiently to make a successful apparatus.

He realised that it would not be possible to obtain any actual magnification of the human voice, but considered that it ought to be possible to obtain from gramophone records sounds in close accord, in respect of tone and loudness, with the original sounds from which the records were made.

It is interesting to note that careful quantitative experiments, which he made at a much later date, showed that this conclusion was correct, and that the energy of the sound waves produced by the loudest air-operated reproducer he was able to construct was no greater than the energy imparted to the needle by the record. He concluded that such magnification as he was able to obtain was due to the fact that the energy required to move the recording or cutting style was less than that delivered by the record to the producing needle, and an air-operated reproducer was very much more efficient, so that the losses between the energy imparted by the record to the needle, and the sound energy produced, were far less than is the case when the energy of the needle is impressed directly on the air column through the medium of a diaphragm; while at the same time the original sound is more correctly reproduced. At the conclusion of his examination of the problem, he laid down in quite definite terms the limits within which he would have to work.

The acceleration forces at the needle of the gramophone when playing loud music at a frequency of 500 periods amounted to 100 times gravity, with a corresponding amplitude of 0·003 inch.

The energy to vibrate a valve at any given frequency varies as the square of the amplitude, and also as the square of the frequency. It also varies as the integral of the mass multiplied by the amplitude of oscillation.

The volume of air passed by a valve having a very narrow lip-opening is much less than is the case with an ordinary well-shaped orifice of reasonable size, viz., the velocity may be only half as great.

The valve must be designed so that the area of opening is approximately proportional to the movement of the valve, and the leakage area, when closed, must be kept negligibly small.

The amplitude of vibration, and weight of all moving parts must be made as small as possible.

The velocity of flow of air through very minute apertures decreases more rapidly than does the opening when dealing with aperture areas of, say, 0·001 inch and under.

PLATE VII

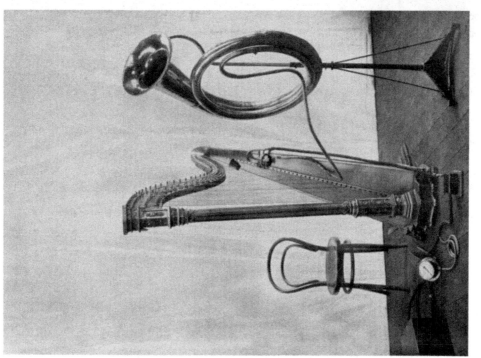

Auxetophone applied to stringed instruments

Assuming operation at an air pressure of 72 inches water column, and at a frequency of 500 cycles, one vibration of sine form, i.e., from nearly closed to open, and back to nearly closed again, will pass from 3 to 5 cubic inches of air per square inch of maximum area of valve opening.

Having thus satisfied himself that he would not be wasting his time, he began to devote the whole of his leisure to the construction of experimental air valves. He did all this work himself in his private workshop at Holeyn Hall, starting directly after dinner and often working far into the following morning. It is interesting to note that he did all this during a period of intense development of the steam turbine, both land and marine, regarding his work on the auxetophone purely as a hobby and a means of relaxation from his ordinary work. Many of his early valves were crudely made from tobacco tins, watch springs, pieces of bone, india-rubber bands, and sealing-wax; but later, satisfied with his progress, he installed a very complete set of fine watchmakers' tools, and produced valves that the most skilled instrument maker would have been proud to show.

When he tried valves of the gridiron slide valve and piston valve types, he found them liable to become stuck or seriously impeded by small particles of dust (even if the air was finely filtered); also, if the fixed and moving faces were given sufficient liberty to ensure free motion, the leakage became so heavy that variations in the opening of the valve were no longer proportional to its movement, with the result that the sound was impaired.

He then made up a valve in which the flat hinged flap moved normally to a flat valve face, in which were formed a double series of parallel rectangular ports, arranged alternately to form inlets and outlets for the air. When the valve flap lifted, air flowed up the inlet ports and discharged over the lips of the adjoining outlet ports and set up a wave of pressure in the trumpet. The moving flap valve was supported on spring trunnions and had attached to it a suitable holder for the needle which ran in the groove of the record and thus caused the valve to oscillate in sympathy with the recorded vibrations. The face of the valve flap was covered with a pad about 0·025 inch thick, composed of alternate layers of the finest gossamer silk and goldbeaters skin, which gave elasticity to the valve face and prevented metallic jarring.

The valve, which he finally adopted, consisted of a wind box containing a cotton-wool and gauze filter through which the air passed on its way to a grid having a number of parallel rectangular openings separated by bars of metal. The valve itself was a flat-faced comb made of magnalium, the tongues of which were spaced so that each tongue just covered (with a very small overlap) one of the rectangular slots in the valve seat. The comb valve was mounted rotationally on a pair of torsional springs so that it moved in a direction substantially normal to the valve seat. The valve was operated indirectly from a second torsionally mounted weighbar which carried the gramophone needle, and the movement of the bar was transmitted to the valve by a push-and-pull rod embedded in a thick viscous fluid contained in holes in both members. This fluid consisted of bicycle tyre cement mixed with sufficient castor oil to give the desired viscosity. This device cut out needle scratch, and made the valve itself independent of any mechanical out-of-truth of the record. The movement of the valve was controlled by a spring operated by a piston in a small cylinder connected to the wind-box. The pressure of the air on the piston was opposed by a screw-operated tension spring, by which the closeness of the vibrating valve to the valve face could be adjusted. The piston also acted to some extent to counterbalance the out-of-balance of the comb valve, caused by the air pressure acting only on one side of it.

The valve was mounted on a right-angle Tee-piece at the end of a tapered tubular arm, pivotted to the horn and fitted with a counterbalance weight to reduce the load on the needle and so avoid undue wear of the record. The Tee-connection introduced a sharp right-angled turn into the sound-path which, with the help of an adjustable hollow plunger packed with cotton-wool, acted as a secondary scratch filter, and also cut out unpleasant harmonics.

In its perfected form, the auxetophone gave results, thirty years ago, which in volume and quality, and in range of operation before cut-off, have only just been approached by the electric gramophone and wireless loud-speakers of the last few years, which owe their amplification to the thermionic valves invented by Sir Ambrose Fleming.

When Parsons applied for patents for his air-valves, he discovered for the first time that a valve on similar principles had been patented by Mr Horace L. Short, who had used it from the top of the Eiffel Tower in Paris. He purchased Short's patent and sold the whole of the gramophone rights outright to the H.M.V. Company. He engaged Short to work at Heaton Works on the further development of the air-valve for use on musical instruments, and applied it successfully to the 'cello and double bass fiddle, the valve being mounted on a wooden bridge clamped on felt pads over the side walls of the fiddle body close to the bridge, but clear of the sound-board or belly of the instrument, so as to avoid interference with its free vibration. The vibration of the sound-board of a fiddle is rotational about the sound-post, the fiddle holes giving freedom for this vibration to take place. The vibration was communicated to the auxetophone valve by means of a fork-ended telescopic rod with viscous fluid to give rigidity of push and pull within the limits of frequency required.

'Cellos and double bass fiddles fitted with auxetophones were in use by Sir Henry Wood in the Queen's Hall Orchestra during the whole of the Winter Season in 1906, and enabled him to achieve a balance of heavy stringed instruments which, owing to space limitation, would have been impossible without them. Although these improved instruments were greatly appreciated by Sir Henry Wood, it proved futile to develop them further owing to the opposition of the musicians, who took the view that one of these instruments could do the work of four or five ordinary ones, and would put a corresponding number of men out of employment.

Parsons also applied the auxetophone to the violin but without much success, as it was found to be too cumbersome.

He also applied it experimentally to the piano and the harp, but these instruments have sound-boards of large area and it was found impossible to select any part of the sound-board which would respond to the vibrations of all the strings. The vibrations appeared to take place in local zones which moved about according to the strings which were being sounded. One "Errard" harp was fitted with four or five separate auxetophone valves, but the results were discouraging.

The auxetophone is perhaps the only one of all Parsons' inventions which may be said to have been ahead of its time. If he had invented it twenty-five years later, it would have been just in time to apply to talking pictures, but, by the time the "talkies" had been developed, the patents had run out and all chance of a monopoly, which would have justified further development, had gone.

In 1922, when wireless broadcasting first came in, the auxetophone was used experimentally as a loud-speaker and gave results much superior to those of any loud-speaker until the present-day moving-coil instruments were developed. Parsons took a keen interest in these experiments but, owing to the patent position, it was no more than sentimental.

Care has been taken to preserve at Heaton Works all the experimental auxetophone valves which Parsons made with his own hands, and which he never even brought to the Works until he wanted valves of the final and successful design produced commercially.

The horn used with the auxetophone on the gramophone consisted of a coned tube of uniform taper about 30 feet long. The small end was about 1 inch diameter and the large end finished off with a bellmouth 3 feet 6 inches diameter. The whole conical horn was worked up into the spiral form shown in Plate VII. Parsons did very little research work on horns themselves, but based his designs on those of the brass horns used in Military bands. Much thought has been put into the design of modern horns and an auxetophone was tried by Sir Charles just before his death on one of the scientifically designed logarithmic horns made by the Western Electric Company for picture-house work, and the musical quality was appreciably improved.

The writer treasures one of the Parsons' valves and maintains it in working order both on the gramophone and on broadcasting.

APPENDIX B

OPTICAL GLASS

At the outbreak of the War there was only one firm in this country producing optical glass, and the danger of supplies of instruments for the Services being cut off in the event of this factory being bombed by aircraft caused the Government such concern that arrangements were made by which a second factory was erected with the aid of financial assistance from the State by a company already interested in other types of glass production. The factory was built just outside Derby at Little Chester.

At the close of the War the company had no further interest in optical glass and desired to dispose of the factory. Sir Charles Parsons who had already taken an interest in optical instrument production was approached by a Government Department, a suggestion being made at the time that there was a possibility of Government financial assistance. He purchased the factory and declined the proffered help, preferring to keep the business entirely in his own hands.

It was no doubt his early association with astronomy which led him to take an interest first in optical instrument production, later in optical glass, and finally in large telescopes.

With his accustomed energy he set about the re-organisation of the Derby factory, rebuilt several of the furnaces and changed over from gas firing to oil, at the same time extending the number and capacity of the annealing lehrs and introducing many improvements in their design and insulation.

Up to this time no special efforts had been made to produce in this factory large glass discs but, when occasion arose, Sir Charles was equal to the task. The firm of Sir Howard Grubb and Sons had on order a large astronomical telescope for the South African Government, but had sought in vain for a pair of discs for the objective. It was known that there was possibly available in the country a disc of crown glass, but there was no suitable negative flint disc. There were suggestions that it might be necessary to obtain a pair of discs from a continental source and he either estimated, or more probably was told, that he could not expect delivery under eighteen months.

A suggestion was made to Sir Howard that it might be possible to obtain the discs from Derby. This resulted in a meeting being arranged between Sir Howard Grubb and Sir Charles, at which the whole question was discussed. Sir Howard laid it down as a condition that he could not wait more than twelve months for the discs or it might result in the cancellation of the

PLATE VIII

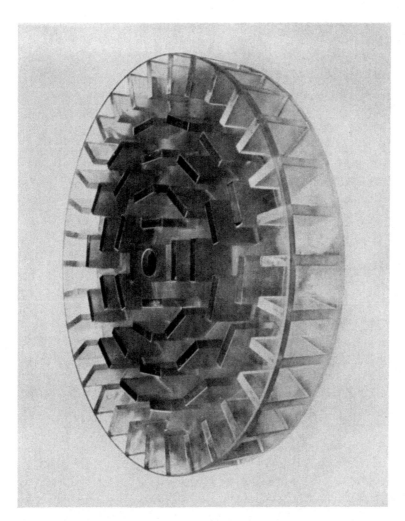

No. 4. Experimental built-up glass mirror

contract. Sir Charles' answer was that which might have been anticipated. He replied, "Very well then, you shall have a pair of discs in nine months". Needless to say he kept his word and that was no mean feat seeing that the works, up to that point, had had no experience in the production of large telescope discs and approximately from a third to half the period would be absorbed in annealing the glass.

After some few preliminary experiments, satisfactory pots of glass were made from which suitable discs were obtained and subsequently annealed. Well within the allotted time the discs were delivered and found to be entirely satisfactory.

His next essay in the production of large discs were two of 42 inches diameter for a foreign government. One of them was shown at the British Empire Exhibition and excited considerable interest.

His success in producing large discs emphasised the belief he had long held that the normal method of producing plates of optical glass was very wasteful.

Probably his greatest advance lay in the production of very large discs for astronomical telescopes by a method which ran counter to all the traditions of the industry existing at that time. After some preliminary experiments he met with complete success, producing discs of a size not previously attempted. The slow cooling of the discs presented a further problem, which was overcome by the erection of a special type of tower having several novel features.

Upon taking over the control of the Derby Crown Glass Works, he concentrated on improving the quality of optical glass and, applying scientific principles, adopted methods which were considered to be revolutionary. These, however, proved successful and achieved his purpose. Some of them are briefly described in the following notes:

Furnace Design. Optical glass is usually melted in pots in single-pot furnaces and, owing to the faulty furnace arrangements, there had been a large percentage of pot breakages involving the loss of whole pots of glass. Parsons therefore re-designed the furnaces to embody top heating which prevented the flames from playing directly on the pot. At the same time the flue was arranged below the centre of the pot, to give a more uniform temperature throughout the furnace. The change was successful and pot breakage practically ceased. Means were also provided for effectively cooling down a melt from the bottom of the pot in order to reduce the introduction of striae due to convection currents and a better form of stirrer was developed.

"Flowing out" Process. The old method of allowing the molten glass to cool out in its pot was very wasteful, as the glass broke up into pieces of all shapes and sizes, and had to be remoulded into blocks. Parsons therefore

set about finding means of producing the glass in one large disc or block and sawing it up as required. The arrangement eventually standardised was to invert the pot of glass and allow the contents to "flow out" into a mould.

After stirring, the glass was partly cooled and the pot removed from the furnace. The pot was then inverted in a special grab and placed upside down on a preheated plate with a moulded ring in position. Special pot-lifting levers were attached, a hole drilled in the pot bottom for air inlet purposes, and the whole put into a preheated furnace of special design. A balanced lifting device which passed through the furnace crown was connected up with the pot levers and the weights adjusted to give a steady upward pull on the pot. The glass, on softening, "flowed out" into the mould. This process was very successful in considerably increasing the yield per pot as large discs, approximately 42 inches diameter by 7 inches thick, were regularly moulded.

Annealing. Perfectly annealed glass is necessary, and to enable large discs to be dealt with satisfactorily special annealing towers were designed, being arranged for both gas and electrical heating. The tower consisted of an insulated canopy carried on suitable lifting gear, and fitted with an internal fan, externally driven, with an air deflecting sleeve to ensure there being a uniform temperature throughout the tower. When cooling-down discs from the "flowing out" process, the discs were run under the canopy and cooled out as required. Ground and polished discs were packed in sand in a container before being put into the tower. The heat was under careful control, and very good annealing was obtained.

Built-up Mirrors, etc. Amongst many other things Sir Charles considered methods of manufacturing large telescope mirrors of a built-up or open construction instead of solid discs, which were difficult to obtain, required heavy and costly mountings, and were subject to temperature variations in service. The built-up arrangement consisted of having two plates of glass separated by a multiplicity of glass pillars—see Plate VIII—the whole being fused together in a specially designed furnace having suitable control and tell-tale gear for regulating the process.

This method was later superseded by a composite pattern of mirror. This consisted of two plates separated by granulated glass, the whole being fused together, giving a mirror with a cellular centre. This pattern of mirror had many advantages and was a comparatively simple type to manufacture.

There were practically one hundred separate types of optical glass manufactured and stocked at Derby and experiments were always being made to obtain special types of glass to meet the requirements of the scientific instrument manufacturers.

It is said that Sir Charles spent upwards of £60,000 in improving the manufacture and quality of optical glass. In June 1925 he purchased the business of Sir Howard Grubb and Sons, Ltd., of St Albans, and established it at Newcastle-upon-Tyne under the title of Sir Howard Grubb Parsons and Co. The new works, adjoining the steam turbine and electrical works at Heaton, were completed in 1926 and shortly afterwards were transferred to C. A. Parsons and Co., Ltd., and now form a subsidiary of the Heaton works.

The factory was especially designed and laid out for the manufacture of large astronomical telescopes and consists of two bays, one comprising the drawing and other offices and optical section, the second forming the light machine and the erecting shop.

APPENDIX C

CONTRIBUTIONS TO SCIENTIFIC AND TECHNICAL LITERATURE

"High Speed Motors." Gisbert Kapp. (Discussion by C. A. Parsons on engines of the compound steam turbine type.) Institution of Civil Engineers, vol. 83 (Paper 2113), p. 263. November 24, 1885.

"The Compound Steam Turbine and its Theory as applied to the Working of Dynamo Electric Machines." C. A. Parsons. N.E. Coast Institution of Engineers and Shipbuilders. *Proceedings*, vol. 4, part 4. December 19, 1887.

"Experiments on Carbon at High Temperatures and under Great Pressures, and in contact with other Substances." C. A. Parsons. Royal Society. *Proceedings A*, vol. 44, p. 320. 1888.

"Alternate Current Machinery." Gisbert Kapp. (Discussion by C. A. Parsons on a bi-polar alternator in which the sine-curve was nearly realised.) Institution of Civil Engineers, vol. 97 (Paper 2392), p. 65. February 19, 1889.

"Description of the Compound Steam Turbine and Turbo-electric Generator." C. A. Parsons. Institution of Mechanical Engineers. *Proceedings*, pp. 480–522. October, 1888.

"Cost of Electrical Energy." R. E. B. Crompton. (Discussion by C. A. Parsons on the working of the Newcastle and District Lighting Station.) Institution of Civil Engineers, vol. 106 (Paper 2498), p. 58. April 7, 1891.

"Application of the Compound Steam Turbine to the purpose of Marine Propulsion." C. A. Parsons. Institution of Naval Architects. *Proceedings*, 38th Session. April 8, 1897.

"The Relative Advantages and Disadvantages of Rotary and Reciprocating Engines as applied to Ship Propulsion." C. A. Parsons (Engineering Conference). Institution of Civil Engineers. *Proceedings*, vol. 130, p. 206. May 26, 1897.

"The Application of the Steam Turbine to the Workings of Dynamos and Alternators." C. A. Parsons (Engineering Conference). Institution of Civil Engineers. *Proceedings*, vol. 130, p. 216. May 26, 1897.

"The Progress of the Steam Turbine." C. A. Parsons. Institute of Marine Engineers, 69th Paper. *Transactions*, vol. 9. October 11 and October 25, 1897.

Presidential Address, 19th Session. C. A. Parsons. Institution of Junior Engineers. *Journal*, vol. 10, part 1. November 3, 1899.

"Motive Power—High-speed Navigation Steam Turbines." C. A. Parsons. The Royal Institution. January 26, 1900. (Sir Frederick Bramwell in the Chair.)

"On the supersession of the Steam by the Electric Locomotive." W. Langdon. (Discussion by C. A. Parsons.) Institution of Electrical Engineers. *Journal*, vol. 30, no. 148, p. 154. November 29, 1900.

"The Marine Steam Turbine and its Application to Fast Vessels." C. A. Parsons. Institution of Engineers and Shipbuilders in Scotland. 50th General Meeting, 44th Session. *Transactions*, vol. 45, part 4. February 19, 1901.

"Trials of Steam Turbines for driving Dynamos." C. A. Parsons and G. G. Stoney. Institution of Mechanical Engineers. *Proceedings*, Section III. September 3, 1901.

"Steam Turbines." C. A. Parsons. Royal Artillery Institution, Woolwich. (Major-General L. W. Parsons, C.B., R.A., in the Chair.) *Proceedings*, vol. 30, nos. 4, 5 and 6. January 29, 1903.

Patents relating to the Auxetophone, 1903–1904: Sound Reproducers or Intensifiers for Phonographs, Gramophones or Telephones, No. 10,468/1903. Musical Instruments, No. 10,469/1903. Reproducers or Resonators, No. 20,892/1904.

"The Steam Turbine and its Application to the Propulsion of Vessels." C. A. Parsons. Institution of Naval Architects. (A. F. Yarrow in the Chair.) *Transactions*. June 26, 1903.

"The Steam Turbine as applied to Electrical Engineering." C. A. Parsons, G. G. Stoney and C. P. Martin. Institution of Electrical Engineers. *Journal*, vol. 33, no. 167, pp. 794–837. May 12, 1904.

"High-speed Electric-Railway Experiments on the Marienfelde-Zossen." A. Siemens. (Discussion by C. A. Parsons.) Institution of Electrical Engineers. *Journal*, vol. 33, no. 168, p. 925. May 26, 1904.

Address to the Engineering Section at the Cambridge Meeting of the British Association. C. A. Parsons. 1904.

Presidential Address. C. A. Parsons. Institute of Mechanical Engineers. January 16, 1905.

"Surface Condensing Plants." R. W. Allen. (Discussion by C. A. Parsons on: (i) Advantage of having separate air and water pumps, p. 217. (ii) Origin of the method of cooling the water entering the air-pumps, p. 217. (iii) Vacuum Intensifier, p. 219. (iv) Condenser Tubes, p. 220. (v) Objection to the arrangement of the Curtis turbine on the condenser, p. 220.) Institution of Civil Engineers. *Proceedings*, vol. 161. February 28, 1905.

"Progress in Methods of Propulsion of Vessels." C. A. Parsons. Navigation Congress, Milan. 1905.

"The Steam Turbine." C. A. Parsons and G. G. Stoney. Institution of Civil Engineers. *Proceedings*, vol. 163, p. 167. December 5, 1905.

"The Steam Turbine on Land and Sea." C. A. Parsons. The Royal Institution. Friday Evening Discourse. May 4, 1906. Reprinted in the *Smithsonian Report* for 1907, pp. 99–112.

"Development of the Marine Steam Turbine." C. A. Parsons and R. J. Walker. (Engineering and Machinery Exhibition.) Institute of Marine Engineers. September 29, 1906.

"Marine Steam Turbine Development." C. A. Parsons and R. J. Walker. N.E. Coast Institution of Engineers and Shipbuilders. *Proceedings*, vol. 23. March 15, 1907.

"Some Practical Points in the Application of the Marine Steam Turbine." C. A. Parsons and H. W. Ridsdale. Bordeaux International Congress. Institution of Naval Architects. June 25, 1907.

"Some Notes on Carbon at High Temperatures and Pressures." C. A. Parsons. The Royal Society. *Proceedings A*, vol. 79, p. 532. 1907.

"The Combination System of Reciprocating Engines and Steam Turbines." C. A. Parsons and R. J. Walker. Institution of Naval Architects. Spring Meetings, 49th Session. April 9, 1908.

"The Conversion of Diamond into Coke in high Vacuum by Cathode Rays." C. A. Parsons and A. A. Campbell Swinton. The Royal Society. *Proceedings A*, vol. 80, p. 184. 1908.

"The Expansive Working of Steam attainable in Steam Turbines." C. A. Parsons. The Greenock Philosophical Society. *Proceedings*. January 15, 1909.

"The Application of the Marine Steam Turbine and Mechanical Gearing to Merchant Ships." C. A. Parsons. Institution of Naval Architects, 51st Session. March 18, 1910.

"Twelve Months' Experience with Geared Turbines in the Cargo Steamer *Vespasian*." C. A. Parsons and R. J. Walker. Institution of Naval Architects, 52nd Session. April 5, 1911.

"Experiments on the Compression of Liquids at High Pressures." C. A. Parsons and S. S. Cook. The Royal Society. *Proceedings A*, vol. 85, p. 332. May 25, 1911.

"The Steam Turbine." C. A. Parsons. Rede Lecture. 1911. Cambridge University Press.

"The Marine Steam Turbine from 1894 to 1910." C. A. Parsons. Institution of Naval Architects. Jubilee Meeting. (Sir William White in the Chair.) *Proceedings*. July 5, 1911.

"Central Heating and Power Plant of McGill University, Montreal." R. J. Durley. (Discussion by C. A. Parsons on steam required for electric generating secondary to that required for heating (p. 243), and advantages of thermal storage (p. 243).) Institution of Civil Engineers, vol. 188 (Paper 3985). January 30, 1912.

"Relative Possibilities of the Diesel Oil Engine, Geared Turbine, and Suction-gas Engine, as compared with the Reciprocating Engine, for Marine Propulsion." E. L. Orde, C. A. Parsons, R. J. Walker and A. C. Holzapfel. N.E. Coast Institution of Engineers and Shipbuilders. *Proceedings* and Reprints. April 1, 1912.

"Mechanical Gearing for the Propulsion of Ships." C. A. Parsons. Institution of Naval Architects. *Proceedings*. March 13, 1913.

Presidential Address. C. A. Parsons. N.E. Coast Institution of Engineers and Shipbuilders. October, 1913.

Address on the Occasion of the Presentation of the Freedom of the City of Newcastle-upon-Tyne to C. A. Parsons. *Proceedings* of the Newcastle Council. 1914.

"The Parallel Operation of Electric Power Stations." J. S. Peck. (Discussion by C. A. Parsons.) Institution of Electrical Engineers. *Journal*, vol. 55, no. 262, p. 74. January 17, 1917.

"Experiments on the Artificial Production of Diamonds." C. A. Parsons. Bakerian Lecture. The Royal Society. April 25, 1918. *Philosophical Transactions*, A, vol. 220, p. 67 (1920).

"The Formation of the Diamond." C. A. Parsons. Institute of Metals. *Proceedings*. 1918.

"Investigations into the Causes of Corrosion or Erosion of Propellers." C. A. Parsons and S. S. Cook. Institution of Naval Architects. *Proceedings*. April 10, 1919.

President's Address. British Association, Bournemouth Meeting. 1919.

"Researches at High Temperatures and Pressures." C. A. Parsons. The Royal Institution, Friday Evening Discourse. January 23, 1920.

"Some Reminiscences of Early Days of Turbine Development." C. A. Parsons. Presentation of the Franklin Medal. May 19, 1920. *Journal Franklin Institute*, July, 1920.

"Reminiscences of Electrical Experiences." C. A. Parsons. Institution of Electrical Engineers. 50th Anniversary Commemoration. *Journal*, vol. 60, no. 308, p. 411. February 23, 1922.

"Motive Power." C. A. Parsons. (Presidential Address.) Birmingham and Midland Institute. *Proceedings*. October 12, 1922.

"The Rise of Motive Power and the Work of Joule." C. A. Parsons. (Joule Memorial Lecture.) Manchester Literary and Philosophical Society. *Proceedings*, vol. 67, part 1. December 5, 1922.

"Mechanical Gearing." C. A. Parsons. Institution of Naval Architects. Spring Meetings, 64th Session. March 22, 1923.

"Physics in Industry." C. A. Parsons. (Presidential Address.) Institute of Physics. March 26, 1924. (Reprinted as a preface to lectures by J. W. Mellor, A. E. Oxley and C. H. Desch.)

"Steam Turbines." C. A. Parsons. World Power Conference at the British Empire Exhibition, Wembley, London. *Proceedings*. July 4, 1924.

"Steam Turbines." C. A. Parsons. International Mathematical Congress of Toronto. August 15, 1924.

"The Steam Turbine—a study in Applied Physics." C. A. Parsons. (1st Foundation Centenary.) Franklin Institute, Pennsylvania. *Journal Franklin Institute*. September 18, 1924.

"Progress in Economy of Turbine Machinery on Land and Sea." C. A. Parsons, R. J. Walker and S. S. Cook. N.E. Coast Institution of Engineers and Shipbuilders. *Proceedings*. January 14, 1927.

"Some Investigations into the Cause of Erosion of the Tubes of Surface Condensers." C. A. Parsons. (Illustrated by a Kinematograph film.) Institution of Naval Architects. April 6, 1927.

"Recent Advances in Steam Turbines." C. A. Parsons. (Read by R. Dowson.) Delft University. February 22, 1928.

"The Problem of Artificial Production of Diamonds." Initialled "C. H. D." Summarises the results of C. A. Parsons, H. M. Duncan and others. *Nature*, vol. 121, No. 3055. May 19, 1928.

"Direct Generation of Alternating Current at High Voltages." C. A. Parsons and J. Rosen. Institution of Electrical Engineers, Spring Meeting. *Journal*, vol. 67, p. 1065, and vol. 68, p. 411. January 15 and October 29, 1929.

"A New Method of Casting Steel Ingots." C. A. Parsons. Iron and Steel Institute. *Journal*. May 5, 1929.

"Recent Progress in Steam Turbine Plant." C. A. Parsons. British Association, South Africa Meeting. July 31, 1929.

"The Future of the Marine Steam Turbine." C. A. Parsons. *The Shipping World*. April 9, 1930.

"Further Development in the Application of High-pressure Steam to Navigation." C. A. Parsons, R. J. Walker and S. S. Cook. World Power Conference, 2nd Plenary Meeting. *Transactions*. June 18, 1930.

"The Use in Power-stations of Steam Turbines having, with their Auxiliaries, large over-load Capacities." C. A. Parsons and R. Dowson. World Power Conference. *Transactions*. June 23, 1930.

INDEX